Lecture Notes in Economics and Mathematical Systems

Managing Editors: M. Beckmann and H. P. Künzi

Mathematical Economics

124

Michael J. Todd

The Computation of Fixed Points and Applications

Springer-Verlag
Berlin · Heidelberg · New York 1976

Author
Michael J. Todd
School of Operations Research
and Industrial Engineering
Cornell University
Ithaca, New York 14853/USA

Library of Congress Cataloging in Publication Data

Todd, Michael J 1947-
The computation of fixed points and applications.

(Lecture notes in economics and mathematical systems ; 124) (Mathematical economics)
Bibliography: p.
Includes index.
1. Fixed point theorems (Topology) 2. Triangulating manifolds. 3. Economics, Mathematical.
I. Title. II. Series. III. Series: Mathematical economics.
QA611.7.T63 514'.3 76-9786

AMS Subject Classifications (1970): 05-01, 55-04, 55C20, 57C15, 90A15, 90C30

ISBN 3-540-07685-9 Springer-Verlag Berlin · Heidelberg · New York
ISBN 0-387-07685-9 Springer-Verlag New York · Heidelberg · Berlin

Printed in Germany
Printing and binding: Beltz Offsetdruck, Hemsbach/Bergstr.

TABLE OF CONTENTS

PREFACE

Fixed-point algorithms have diverse applications in economics, optimization, game theory and the numerical solution of boundary-value problems. Since Scarf's pioneering work [56,57] on obtaining approximate fixed points of continuous mappings, a great deal of research has been done in extending the applicability and improving the efficiency of fixed-point methods. Much of this work is available only in research papers, although Scarf's book [58] gives a remarkably clear exposition of the power of fixed-point methods. However, the algorithms described by Scarf have been superseded by the more sophisticated restart and homotopy techniques of Merrill [48,49] and Eaves and Saigal [14,16]. To understand the more efficient algorithms one must become familiar with the notions of triangulation and simplicial approximation, whereas Scarf stresses the concept of primitive set.

These notes are intended to introduce to a wider audience the most recent fixed-point methods and their applications. Our approach is therefore via triangulations. For this reason, Scarf is cited less in this manuscript than his contributions would otherwise warrant. We have also confined our treatment of applications to the computation of economic equilibria and the solution of optimization problems. Hansen and Koopmans [28] apply fixed-point methods to the computation of an invariant optimal capital stock in an economic growth model. Applications to game theory are discussed in Scarf [56,58], Shapley [59], and Garcia, Lemke and Luethi [24]. Allgower [1] and Jeppson [31] use fixed-point algorithms to find many solutions to boundary-value problems. Infinite-dimensional cases are also discussed by Freidenfelds [21] and Wilmuth [73]. The Schauder Projection theorem (see Freidenfelds [21] or Smart [61]) describes how a finite-dimensional approximation can be obtained. Our treatment is confined to finite-dimensional spaces throughout. More recent developments we have been unable to cover are the orientation theories of Shapley [60], Lemke and Grotzinger [44], Todd [66] and most generally Eaves [15] and Eaves and Scarf [17]; and the algorithmic improvements and convergence analysis of Saigal [53,54]. Eaves [15] contains a comprehensive bibliography.

The manuscript is organized as follows. Chapter I gives a classical (non-algorithmic) proof of Brouwer's theorem from Sperner's lemma, thus introducing the reader to several important concepts. A number of applications are described in Chapter II. We provide a formal treatment of triangulations in Chapter III with descriptions of some important particular triangulations. The latter are used in Chapter IV in algorithms for computing approximate fixed points of continuous functions. The applications of Chapter II motivate extensions from functions to point-to-set mappings. Chapters V and VI parallel Chapters I and II in proving and applying Kakutani's fixed-point theorem. Chapter VII describes an algorithm for computing Kakutani fixed points. This algorithm and those of Chapter IV are inefficient if a good approximation is desired. In Chapters VIII and IX we describe the more sophisticated restart and homotopy algorithms. The latter require special triangulations which are developed in Chapter X. Finally, Chapter XI describes some measures that can be used to compare different triangulations when used for fixed-point computation.

We have included a number of challenging exercises to increase the reader's understanding of the material. Some numerical examples of the algorithms are given in the text. The reader is assumed to be familiar with real analysis and linear programming, including lexicographic resolution of degeneracy. We also assume the Kuhn-Tucker conditions for nonlinear programming known.

This manuscript arose out of a course in computing fixed points that I gave at Cornell University in spring 1975. I am grateful to Michel Cosnard, Pierre Déjax, Pradeep Dubey, Etienne Loute, Shigeo Muto, Bob Rovinsky and Prakash Shenoy for preparing excellent notes. The National Science Foundation, through grant GK-42092, provided support during the preparation of this manuscript. I would like to thank Mrs. Kathy King for her excellent typing. Finally, my thanks go to my wife Marina for her encouragement and assistance.

Notation

N: Set of integers $\{1,2,\ldots,n\}$.

N_0: Set of integers $\{0,1,\ldots,n\}$.

R^m: Set of m-dimensional real column vectors, with coordinates generally indexed 1 through m. However, the coordinates of R^{n+1} are indexed by N_0.

u^i: ith unit vector in R^n, $i \in N$; $u = \sum_{i \in N} u^i$.

v^j: jth unit vector in R^{n+1}, $j \in N_0$; $v = \sum_{j \in N_0} v^j$.

R^m_+: Nonnegative orthant of R^m; $\{x \in R^m | x \geq 0\}$.

$C \cup D$, $C \cap D$, $C \sim D$: Union, intersection and set difference of the sets C and D.

$C + D$, $C - D$: Algebraic sum and difference of sets C and D in R^m;

$\{c + d | c \in C, d \in D\}$ and $\{c - d | c \in C, d \in D\}$ respectively.

λC: $\{\lambda c | c \in C\}$.

$\|x\|_2$: Euclidean norm of the vector $x \in R^m$; $(\sum_i x_i^2)^{1/2}$.

$\|x\|_\infty$: ℓ_∞ norm of $x \in R^m$; $\max_i |x_i|$.

$\|A\|_p$: The ℓ_p operator norm; $\max\{\|Ax\|_p \,|\, \|x\|_p = 1\}$ for $p = 2, \infty$ where A is a real $k \times m$ matrix.

B^m: Euclidean unit ball in R^m; $\{x \in R^m | \|x\|_2 \leq 1\}$.

$B(x,\rho)$: Ball with center x radius ρ; $\{x\} + \rho B^m$ if $x \in R^m$.

$B(C,\rho)$: $C + \rho B^m$ if $C \subseteq R^m$.

$\overline{C}$: Closure of C; $\cap \{B(C,\varepsilon) | \varepsilon > 0\}$.

int C: Interior of C; $\{x \in C | \exists\, \varepsilon > 0$ with $B(x,\varepsilon) \subseteq C\}$.

$\text{diam}_p C$: Diameter of C; $\sup\{\|x-y\|_p | x,y \in C\}$, $p = 2$ or ∞.

$\text{mesh}_p G$: Mesh of G; $\sup\{\text{diam}_p C | C \in G\}$ for $p = 2$ or ∞, where G is a family of subsets of R^m.

CHAPTER I: BROUWER'S THEOREM

I.1. Probably the most famous fixed-point theorem states that a continuous function from a closed n-cell to itself leaves at least one point fixed. The Dutch mathematician L. E. J. Brouwer proved this result in 1912 [5] using degree theory. An equivalent theorem in the case of differentiable functions was proved earlier by Bohl [4], who used Green's theorem. Our concern is more with the computation of fixed points than their existence; we will follow a later approach based on the purely combinatorial lemma of Sperner [62]. This approach is closest to the algorithms we shall develop, and the machinery will be of value later. However, to avoid some of the cumbersome details, we will give only an intuitive idea of the notions of simplex and triangulation. A rigorous treatment appears in Chapter III.

In this section we state Brouwer's theorem formally. Section 2 shows that a proof for the standard simplex is sufficient and gives some examples suggesting various methods of proof. In Section 3 we reduce Brouwer's theorem to Sperner's lemma, which is proved in Section 4.

1.1 Definition. A function h is a homeomorphism if it is one-to-one and onto, and both h and h^{-1} are continuous. A closed n-cell is a homeomorphic image of B^n, i.e., C is a closed n-cell if there is a homeomorphism $h: B^n \to C$.

1.2 Theorem (Brouwer). Let C be a closed n-cell and let $f: C \to C$ be continuous. Then f has a fixed point, i.e., there is an $x^* \in C$ with $f(x^*) = x^*$.

I.2. In this section we show that is is sufficient to prove 1.2 when C is the standard simplex and give some examples of the necessity of the conditions and possible methods of proof.

2.1 Definition. The standard simplex S^n is the convex hull of $v^0, v^1, \ldots, v^n$ in R^n, i.e., $S^n = \{x \in R_+^{n+1} | v^T x = 1\}$. For $i \in N_0$, S_i^n denotes the face of S^n opposite v^i, i.e., $\{x \in S^n | x_i = 0\}$, and the boundary of S^n is $\partial S^n = \cup_{i \in N_0} S_i^n$.

We show below that S^n is a closed n-cell, but to get an intuitive feel for this concept, we first give an easily visualized subclass:

2.2 Lemma. If $C \subseteq R^n$ is compact and convex with a nonempty interior, then C is a closed n-cell.

Proof. We construct a function $h: B^n \to C$ and show that it is one-to-one and onto. The proof that h and h^{-1} are continuous is omitted.

Pick $c \in \text{int } C$. For $0 \neq d \in R^n$ define $\theta(d) = \max\{\theta \in R | c + \theta d \in C\}$. (The maximum is attained because C is compact.) Since $c \in \text{int } C$, $\theta(d) > 0$, and $\theta(\lambda d) = \lambda^{-1}\theta(d)$ for $\lambda > 0$. (θ is closely related to the Minkowski functional; see [61].)

Now define $h: B^n \to C$ by setting $h(d) = c + \|d\|_2\theta(d)d$ if $d \neq 0$, and $h(0) = c$. Clearly, $h(d) = c$ iff $d = 0$, so that $h(d) = h(d') = c \Rightarrow d = d'$. Now let $h(d) = h(d') \neq c$; then $\|d\|_2\theta(d)d = \|d'\|_2\theta(d')d'$ so that $d' = \lambda d$, $\lambda > 0$. But then $\|d\|_2\theta(d)d = \lambda\|d\|_2\lambda^{-1}\theta(d)\lambda d$, and $\lambda = 1$. Thus h is one-to-one.

We now show that h is onto. Since $h(0) = c$, we only need to find $d \in B^n$ so that $h(d) = x$ for $x \in C$, $x \neq c$. Let $d = (x-c)/\|x-c\|_2\theta(x-c)$. It is easily checked that $d \in B^n$ and $h(d) = x$. □

We use 2.2 to show that S^n is a closed n-cell by introducing the set $C^n = \{x \in R^n | 1 \geq x_1 \geq \ldots \geq x_n \geq 0\}$, which will appear in Chapter III.

2.3 Lemma. S^n is a closed n-cell.

Proof. Since by 2.2 C^n is a closed n-cell, we need only show that S^n and C^n are homeomorphic. Define $h: C^n \to S^n$ and $h^{-1}: S^n \to C^n$ by $h(c) = v^0 + Qc$, $h^{-1}(s) = Q's$ with Q the $(n+1)\times n$ matrix $\begin{bmatrix} -1 & & 0 \\ +1 & \ddots & \\ & \ddots & -1 \\ 0 & & +1 \end{bmatrix}$ and Q' the $n\times(n+1)$ matrix $\begin{bmatrix} 0 & 1 & \cdots & 1 \\ \vdots & \ddots & & \vdots \\ \vdots & & \ddots & \vdots \\ 0 & \cdots & 0 & 1 \end{bmatrix}$. Obviously, h is one-to-one and onto, and h and h^{-1} are both continuous. Thus S^n is homeomorphic to C^n and hence to B^n. □

The following result shows that it is sufficient to prove Brouwer's theorem for S^n. While this simplifies the proof, it is certainly not helpful when one wants to compute a fixed point of some other closed n-cell, as one needs explicitly a homeomorphism between the two. This problem will be satisfactorily dealt with when we treat Kakutani's theorem in Chapter V.

2.4 Lemma. Brouwer's theorem is true for all closed n-cells C if it is true for S^n.

Proof. Let $f: C \to C$ be continuous. We must show that f has a fixed point. Since C and S^n are closed n-cells, we have homeomorphisms $h: B^n \to C$ and $h_0: B^n \to S^n$. Then the composite function $f_0 = h_0 h^{-1} f h h_0^{-1}: S^n \to S^n$ is continuous and has a fixed point x^* by hypothesis. Hence $f(hh_0^{-1}(x^*)) = hh_0^{-1}(x^*)$ and f has a fixed point. □

2.5 Examples of the necessity of the conditions of 1.2.

(i) C is not a closed n-cell (see also exercise 5.1).

(a) C is closed and $\text{int } C \neq \emptyset$, but C is not convex. Let $C = \overline{2B^2 \sim B^2}$ and $f(x) = -x$. Then f rotates the torus by π and is clearly continuous, but has no fixed points.

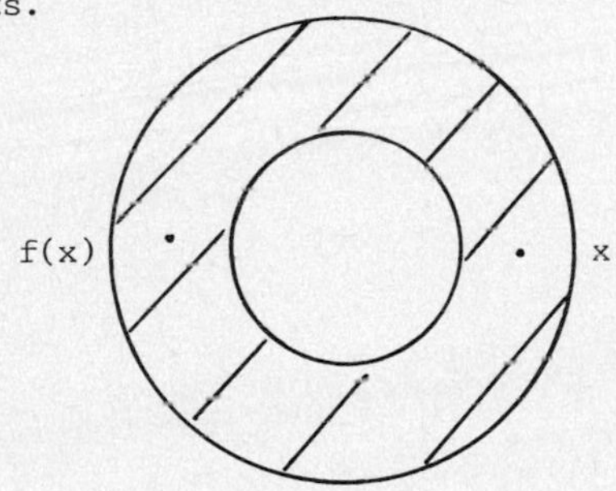

(b) C is convex and $\text{int } C \neq \emptyset$, but C is not closed. Let $C = \text{int } B^2$ and $f(x) = \frac{1}{2} x + \frac{1}{2} u^1$. Clearly f is continuous and has no fixed points.

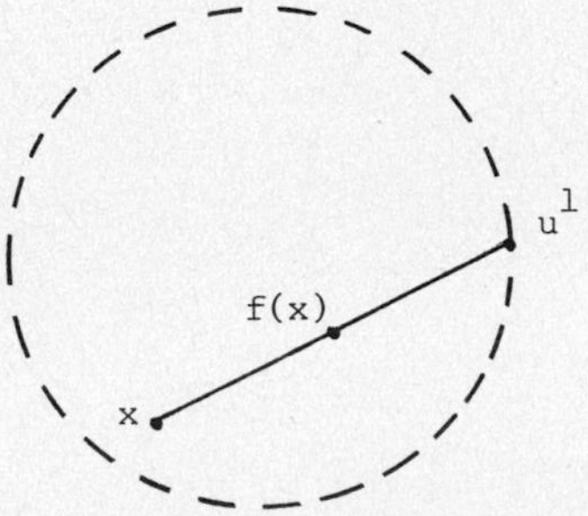

(ii) f is not continuous.

Let $C = B^2$ and $f(x) = \begin{cases} 0 & \text{if } x \neq 0 \\ u^1 & \text{if } x = 0 \end{cases}$. Then f has no fixed points.

2.6 Possible Methods of Proof.

(a) (n = 1) Since S^1 is homeomorphic to [0,1], consider a continuous function $f: [0,1] \to [0,1]$. If $f(0) = 0$ or $f(1) = 1$, we are done. Otherwise, let $g(x) = f(x) - x$; then g is continuous with $g(0) > 0 > g(1)$. By the intermediate value theorem, g has a zero x^* in [0,1] and hence x^* is a fixed point of f. Intuitively, the graph of f must cross the diagonal of the unit square, giving a fixed point:

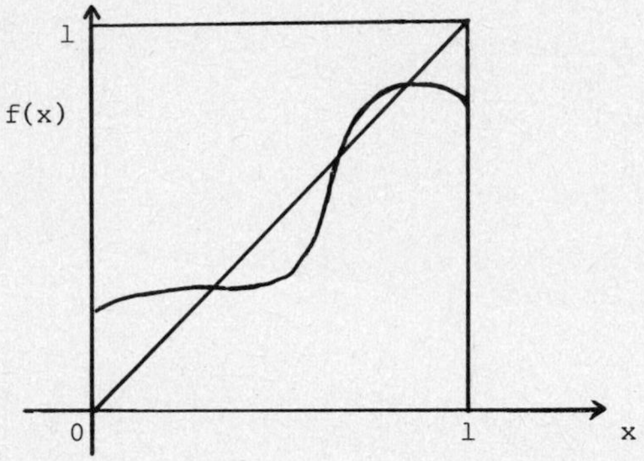

It is not clear how the argument can be extended to higher dimensions--but see (c).

(b) (n = 2) Let us assume we have a continuous function $f: B^2 \to B^2$ without fixed points and try to establish a contradiction. We can then define a function $h: B^2 \to \partial B^2 \equiv \{x \in R^2 \mid \|x\|_2 = 1\}$ by setting h(x) to be the point of ∂B^2 on the line from f(x) to x:

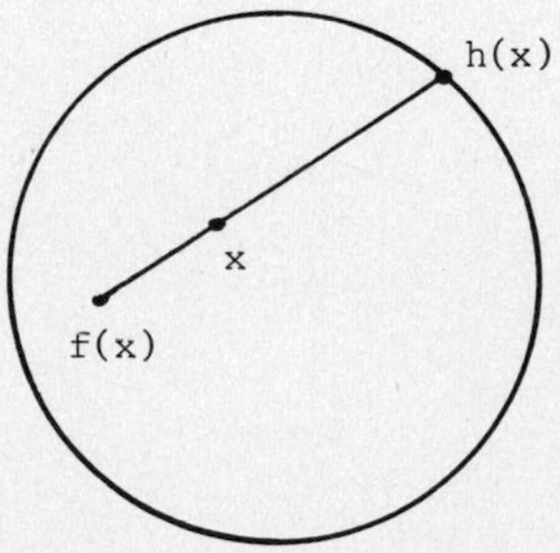

If f has no fixed points, it can be shown that h is continuous, and clearly h leaves fixed each point of ∂B^2. It seems intuitively clear that no function can carry B^2 into ∂B^2, leaving the latter fixed, without "ripping" the interior of B^2 and thus losing its continuity. One can make this argument rigorous for any n, using either homology or the more elementary proof of Hirsch [30]. Hirsch in effect traces the inverse image $h^{-1}(b)$ of a point $b \in \partial B^2$ and shows that it can only "disappear" inside B^2. It must disappear near a fixed point of f; Kellogg, Li, and Yorke [35] have recently constructed an algorithm based on this idea--they require f to be twice continuously differentiable. We will not pursue this approach.

(c) We now motivate the method we will follow. Consider the case $n = 1$. Let $f: S^1 \to S^1$ be continuous. For $i = 0,1$, let C_i be the set of points of S^1 whose i^{th} coordinate does not increase under f. Clearly, each point of S^1 lies in either C_0 or C_1; $v^0 \in C_0$ and $v^1 \in C_1$; and C_0 and C_1 are closed. If C_0 and C_1 intersect, then any point in their intersection must be a fixed point of f, for neither of its coordinates can increase under f while their sum remains 1. Thus the first step in our proof of Brouwer's theorem is to reduce it to the lemma of Knaster, Kuratowski and Mazurkiewicz, stating that under certain conditions a family of sets must have a nonempty intersection. Unfortunately, this lemma seems no easier to prove than the original theorem, and we will have to reduce it further to a purely combinatorial lemma. Consider again the case $n = 1$. If we can find arbitrarily close pairs of points, one in C_0 and one in C_1, then since C_0 and C_1 are closed and S^1 is compact, our lemma will be established. To provide an abundant supply of pairs of points, we can divide S^1 into a family of small segments. An endpoint of a segment will be labelled 0 or 1 according to whether it lies in C_0 or C_1. (If it lies in both, the lemma is proved.) Since one end of S^1 has the label 0 and the other the label 1, it is clear that there is a small segment whose endpoints are labelled 0,1, thus yielding a close pair of points. We have the following picture:

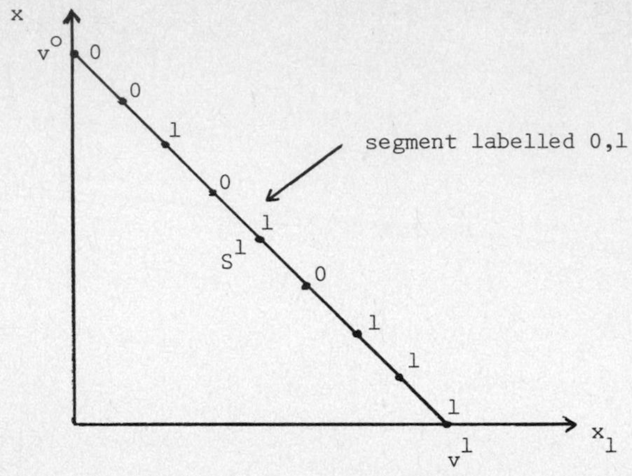

In higher dimensions we must find (n+1)-tuples of close points, one in each C_i. Again we obtain them by dividing S^n into small pieces, called simplices. We divide S^2 into triangles, S^3 into tetrahedra, and so on. In this way we reduce our lemma about families of closed sets to Sperner's lemma concerning labelled simplicial subdivisions.

I.3. In this section we reduce Brouwer's theorem for S^n to Sperner's lemma. Although this step can be made directly (3.7), we proceed more slowly (and, we hope, intuitively) via the Knaster-Kuratowski-Mazurkiewicz (K-K-M) lemma [36]. (This lemma has other interesting applications; see exercise 5.3.)

3.1 Lemma (K-K-M). Let C_i, $i \in N_0$, be a family of closed subsets of S^n satisfying the following conditions:

(i) $S^n = \bigcup_{i \in N_0} C_i$; and

(ii) If $\emptyset \neq I \subseteq N_0$ and $J = N_0 \sim I$, then $\bigcap_{i \in I} S_i^n \subseteq \bigcup_{j \in J} C_j$.

Then $\bigcap_{i \in N_0} C_i \neq \emptyset$.

For $n = 1$, (ii) says only that $v^0 \in C_0$ and $v^1 \in C_1$; for $n = 2$, the conditions are illustrated by the picture below. The shaded section shows that $\bigcap_{i \in N_0} C_i \neq \emptyset$.

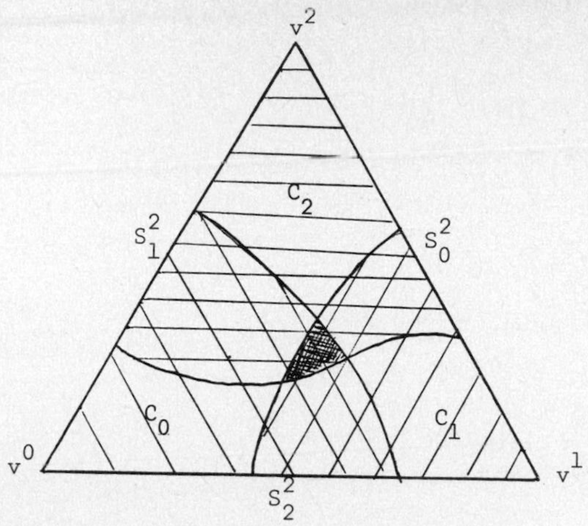

3.2 The K-K-M Lemma implies Brouwer's Theorem for S^n.

Proof. Let $f: S^n \to S^n$ be continuous. For $i \in N_0$ let $C_i' = \{x \in S^n | f_i(x) \leq x_i > 0\}$ and $C_i = \overline{C_i'}$. We now show that the C_i, $i \in N_0$, satisfy the hypotheses of the K-K-M lemma. If $x \in S^n$ lies in none of the C_i', $i \in N_0$, then $f_i(x) > x_i$ for $i \in I = \{i \in N_0 | x_i > 0\}$. Then $1 = v^T x = \sum_I x_i < \sum_I f_i(x) \leq v^T f(x) = 1$, a contradiction. Hence $S^n \subseteq \bigcup_{i \in N_0} C_i' \subseteq \bigcup_{i \in N_0} C_i$. Since for $i \in N_0$, $C_i \subseteq S^n$, condition (i) is verified. Also, by definition, if $x \in \bigcap_I S_i^n$, $x \notin C_i'$, $i \in I$. Thus for $J = N_0 \sim I$, $x \in \bigcup_{j \in J} C_j' \subseteq \bigcup_{j \in J} C_j$, and condition (ii) is verified.

If the K-K-M lemma holds, we can deduce the existence of $x^* \in \bigcap_{i \in N_0} C_i$. Since f is continuous and $x^* \in \overline{C_i'}$, $i \in N_0$, we have $f_i(x^*) \leq x_i^*$, $i \in N_0$. Now $1 = v^T x^* = v^T f(x^*)$ shows that $f_i(x^*) = x_i^*$, $i \in N_0$, and so x^* is a fixed point of f. □

Note: Scarf [58] has proved a lemma similar to 3.1 in which condition (ii) is replaced by:

(ii)' For all $i \in N_0$, $S_i^n \subseteq C_i$.

One can easily verify that this version also implies Brouwer's theorem for S^n. Freidenfelds [22] has shown that the two versions are equivalent.

3.3 Simplices and Triangulations. As we discussed in 2.6(c), we shall prove 3.1 by exhibiting, for any $\varepsilon > 0$, an (n+1)-tuple of points of S^n within ε of one another, with one point in each C_i. We produce these (n+1)-tuples by dividing S^n into many small simplices. The concepts of simplex and triangulation will be introduced informally now; we postpone rigorous definitions until Chapter III.

A closed j-simplex is the convex hull of j+1 points in general position (affinely independent) in R^k; the points are called the vertices of the simplex. An open j-simplex is the relative interior of a closed j-simplex.

n	Closed n-simplex		Open n-simplex
0	•	(point)	•
1	•———————•	(line segment)	)———————(
2	△	(triangle)	△
3	△	(tetrahedron)	△

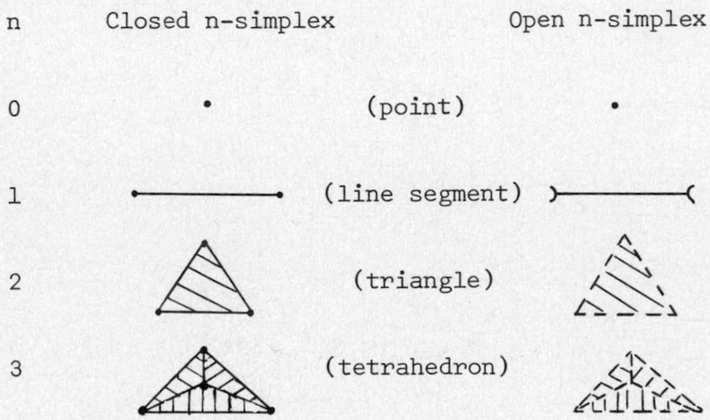

Note: S^n is a closed n-simplex.

A face of a simplex σ is a simplex all of whose vertices are vertices of σ. Thus S_i^n is a closed face of S^n for each $i \in N_0$.

Two simplices are incident if one is a face of the other. Two j-simplices are adjacent if they share a (j-1)-simplex as a face.

A triangulation G of S^n is a finite collection of open n-simplices such that the open n-simplices, together with all their open faces, form a partition of S^n, i.e., S^n is their disjoint union. As we shall see in Chapter III, this rather non-intuitive definition, when couched in terms of closed n-simplices, is equivalent to the conditions:

(i) the closed n-simplices cover S^n; and

(ii) if two closed n-simplices meet, their intersection is a common face.

Thus the following configurations are ruled out:

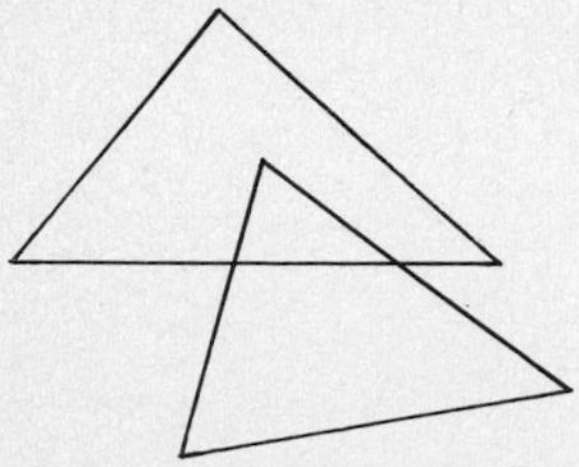

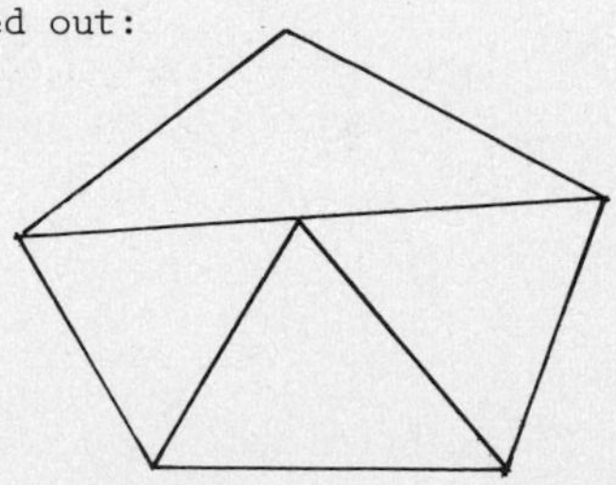

Examples for n = 2.

An irregular triangulation:

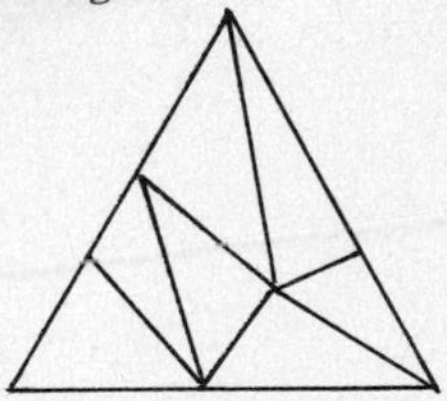

A regular triangulation:

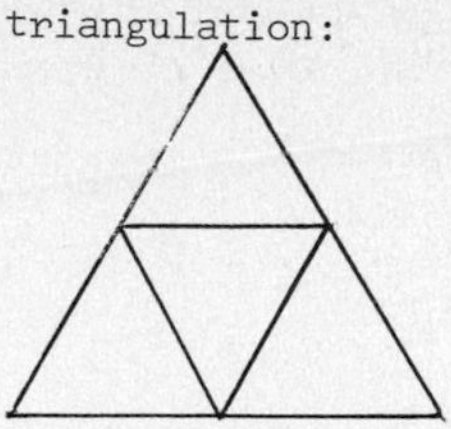

Exercise 5.5 shows that the construction of regular triangulations for $n > 2$ is not as easy as it appears for $n = 1$ and $n = 2$.

We call a vertex of a simplex of a triangulation G a vertex of G. Henceforth, a simplex will mean an open simplex, but little confusion will result from thinking of closed simplices.

3.4 Properties of Triangulations of S^n. The following intuitive results will be proved in Chapter III.

(a) If G is a triangulation of S^n and τ is an (n-1)-simplex that is a face of an n-simplex of G, then either

(i) $\tau \subseteq \partial S^n$ and τ is a face of just one $\sigma \in G$; or

(ii) $\tau \not\subseteq \partial S^n$ and τ is a face of precisely two simplices in G.

(b) There exist triangulations of S^n of arbitrarily small mesh. (Recall that the mesh of G is $\sup_{\sigma\in G}$ diam σ.)

(c) Let G be a triangulation of S^n and $i \in N_0$. Then the collection G' of (n-1)-simplices that are faces of simplices of G and lie in S^n_i is a triangulation of S^n_i (defined in the obvious way).

We can now state

3.5 Lemma (Sperner [62]). Let G be a triangulation of S^n with each vertex of G labelled with an integer in N_0 such that no vertex in S^n_i is labelled i. (Such a labelling is called admissible.) Then there is a simplex in G whose vertices carry all the labels in N_0 (a completely-labelled simplex).

Example for $n = 2$.

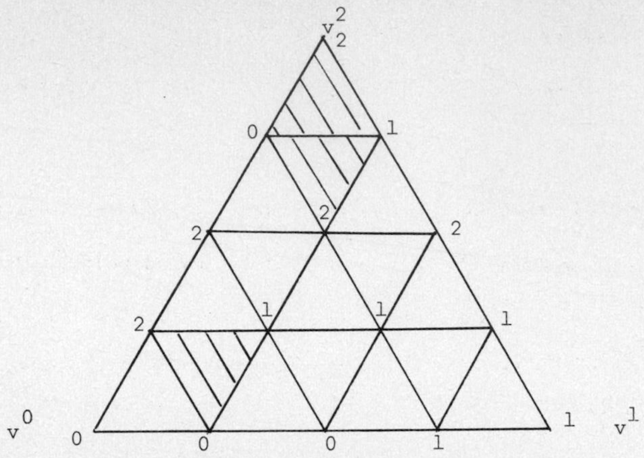

The statement above is the weak form. The strong form of the lemma asserts the existence of an odd number of completely-labelled simplices. (There is even a "super-strong" version. In the example above, note that there is one more small triangle with the labels 0, 1, 2 appearing counter-clockwise (the same direction as v^0, v^1 and v^2 in S^2) then in the reverse direction. The super-strong version asserts this in general, for any n, but the statement and its proof require orientation arguments.)

For related combinatorial results, see Kuhn [37], Tucker [69], whose lemma relates to antipodal fixed-point theorems, and Ky Fan [18], who synthesized Sperner's and Tucker's lemmas in very general results.

3.6 Sperner's Lemma (weak form) implies the K-K-M Lemma.

Proof. Let C_i, $i \in N_0$, be closed sets satisfying the hypotheses of the K-K-M lemma. Using 3.4(b), let G_k, $k = 1,2,\ldots$, be a sequence of triangulations of S^n with mesh $G_k \to 0$. For each k, label each vertex y of G with $i = \min\{j \in N_0 | y \in C_j,\ y \notin S^n_j\}$. (The existence of i follows from the conditions (i)-(ii) of 3.1.) This is an admissible labelling; if Sperner's lemma is true, we can deduce the existence of a completely-labelled simplex σ_k in G_k. Let the vertices of σ_k be y^{ki}, $i \in N_0$, with y^{ki} labelled i, so that $y^{ki} \in C_i$, $i \in N_0$.

Now the sequence y^{k0}, $k = 1,2,\ldots$, lies in the compact set S^n, and therefore there is a convergent subsequence. Without loss in generality, assume

$y^{k0} \to x^* \in S^n$. Since mesh $G_k \to 0$, we have $\lim_{k\to\infty} y^{ki} = x^*$ for each $i \in N_0$. Since C_i is closed, we deduce that $x^* \in C_i$ for all $i \in N_0$, which establishes the K-K-M lemma. □

One might conclude that when Sperner's lemma is used to prove Brouwer's theorem, a completely-labelled simplex gives an approximate fixed point. The result below makes this precise. It is important to note that we use the term "approximate fixed point" to mean a point that is close to its image, while not necessarily close to a fixed point.

3.7 Lemma. Let G be a triangulation of S^n of mesh_∞ at most δ. Let $f: S^n \to S^n$ be such that $\|x-z\|_\infty < \delta$ implies $\|f(x) - f(z)\|_\infty \leq \varepsilon$. Label a vertex y of G $i = \min\{j \mid f_j(x) \leq x_j > 0\}$. Then if σ is a completely-labelled simplex of G and $x^* \in \sigma$, $\|f(x^*) - x^*\|_\infty \leq n(\varepsilon+\delta)$.

Proof. Let σ have vertices y^i, $i \in N_0$, where y^i has label i. Then for each $i \in N_0$ we have

$$f_i(x^*) - x_i^* = (f_i(x^*) - f_i(y^i)) + (f_i(y^i) - y_i^i) + (y_i^i - x_i^*).$$

The hypotheses of the lemma guarantee that the first term on the right-hand side is at most ε, while the last is at most δ. Since y^i has label i, the middle term is nonpositive. Hence for each $i \in N_0$,

$$f_i(x^*) - x_i^* \leq \varepsilon + \delta.$$

Also, since $v^T f(x^*) = v^T x^* = 1$, we have for each $i \in N_0$

$$f_i(x^*) - x_i^* = -\textstyle\sum_{j\neq i} (f_j(x^*) - x_j^*) \geq -n(\varepsilon+\delta).$$

The conclusion now follows. □

Exercise 5.6 asks the reader to develop an admissible labelling scheme that allows the factor n above to be eliminated. If enough is known about the function f, 3.7 allows one to find a point within γ of its image. Let $\varepsilon = \gamma/2n$ and choose δ so that δ and ε satisfy the condition in 3.7. Then triangulate S^n with a triangulation of mesh at most $\min\{\delta,\varepsilon\}$ and find a completely-labelled simplex (by exhaustive search or by one of the algorithms of Chapter IV).

I.4 Proof of Sperner's Lemma. See Lyusternik [41] for an intuitive proof for $n = 2$. We will prove the strong form of Sperner's lemma by induction on n. We could start with the case $n = 0$, which is trivial. However, we start with $n = 1$; we also forgo some easy proofs of this case and instead use the same line of reasoning we employ for the inductive step.

4.1 The Case $n = 1$. The picture is as follows:

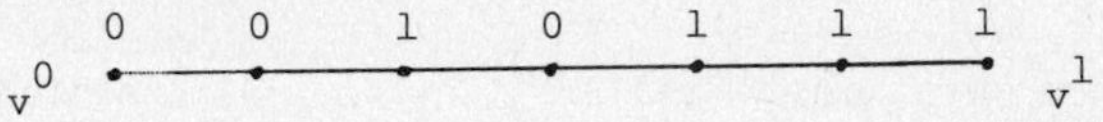

Consider the incidences of vertices ((n-1)-simplices) labelled 0 with segments (n-simplices) of the triangulation. We will count these incidences in two ways.

(i) Add up the incidences for each segment with at least one endpoint labelled 0. Such segments split into two sets:

A - the set of segments with one endpoint labelled 0 and the other 1; and

B - the set of segments with both endpoints labelled 0.

Each segment in A contributes 1 to the count of incidences, and each segment in B contributes 2. Hence the total count is $|A| + 2|B|$.

(ii) Add up the incidences for each vertex labelled 0. Such vertices also split into two sets:

C - the set consisting of just v^0, the only vertex in ∂S^1 labelled 0; and

D - the set of internal vertices (not in ∂S^1) labelled 0.

Using 3.4(a) (trivial in this case), we see that the vertex in C contributes 1 while each vertex in D contributes 2 to the count of incidences. So the total count is $|C| + 2|D| = 1 + 2|D|$.

From $|A| + 2|B| = 1 + 2|D|$ we deduce that $|A|$ is odd, which is precisely the claim of Sperner's lemma for $n = 1$.

4.2 The Inductive Step. Assume that the strong form of the lemma is true for dimension $n-1$, and let G be a triangulation of S^n with the vertices of G admissibly labelled. Let H be the collection of (n-1)-simplices that are faces of simplices in G and whose vertices carry all the labels $0,1,\ldots,n-1$. We will again count the incidences of G and H.

(i) Add up the incidences for each $\sigma \in G$ whose vertices have the labels $0,1,\ldots,n-1$. Such simplices fall into two sets:

A - the set of completely-labelled simplices of G; and

B - the set of simplices of G whose vertices carry all the labels $0,1,\ldots,n-1$ but not n.

Each simplex in A has just one face in H, while each simplex in B has precisely one duplicated label among its vertices, and hence two faces in H--namely, those faces opposite the two vertices sharing the same label. Thus the total count of incidences is $|A| + 2|B|$.

(ii) Add up the incidences for each $\tau \in H$. H is partitioned into two sets:

C - the set of $\tau \in H$ with $\tau \subseteq \partial S^n$; and

D - the set of $\tau \in H$ with $\tau \not\subseteq \partial S^n$.

By the rules of an admissible labelling, each $\tau \in C$ must lie in S^n_n.

By 3.4(a), each $\tau \in C$ contributes 1 and each $\tau \in D$ contributes 2 to the count of incidences of G and H. Thus the total count is $|C| + 2|D|$ and we obtain $|A| + 2|B| = |C| + 2|D|$.

To show that $|A|$ is odd, we must show that $|C|$ is odd. But consider the collection G' of (n-1)-simplices that are faces of simplices of G and lie in S^n_n. By 3.4(c), G' triangulates S^n_n. Moreover, S^n_n is clearly homeomorphic to S^{n-1}, by the deletion or addition of the last zero coordinate. Considered as a triangulation of S^{n-1}, the vertices of G' are admissibly labelled, and the inductive hypothesis allows us to claim that there is an odd number of completely-labelled simplices in G' (i.e., simplices whose vertices carry all the labels $0,1,\ldots,n-1$). But these simplices are precisely those belonging to C; hence $|C|$ and $|A|$ are odd, and the inductive step is complete. $\square$

From 2.4, 3.2, and 3.6, we have finally proved Brouwer's theorem.

4.3 Example ($n = 2$). We show below a labelled triangulation of S^2 and identify the sets A, B, C, and D and their incidences.

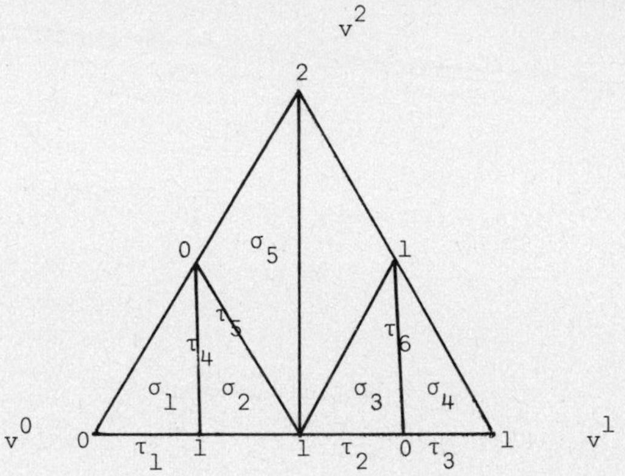

2-simplices in A	Incident 1-simplices in C ∪ D
σ_5	τ_5
2-simplices in B	
σ_1	τ_1, τ_4
σ_2	τ_4, τ_5
σ_3	τ_2, τ_6
σ_4	τ_3, τ_6

Total incidences = $|A| + 2|B| = 9$

1-simplices in C	Incident 2-simplices in A ∪ B
τ_1	σ_1
τ_2	σ_3
τ_3	σ_4
1-simplices in D	
τ_4	σ_1, σ_2
τ_5	σ_2, σ_5
τ_6	σ_3, σ_4

Total incidences = $|C| + 2|D| = 9.$

Note that the simplices in $A \cup B$ form paths: $\sigma_1 - \sigma_2 - \sigma_5$ and $\sigma_3 - \sigma_4$. These paths will be the basis of our algorithms in Chapter IV.

I.5 Exercises

5.1. Show how to construct, for any $C \subseteq R^n$ that is convex but not compact, a continuous function $f: C \to C$ without fixed points.

5.2 [69]. Let $g: B^n \to R^n$ be continuous. Show that either

(i) there is an $x^* \in B^n$ with $g(x^*) = 0$; or

(ii) there are $y^*, z^* \in \partial B^n = \{x \in R^n \mid \|x\| = 1\}$ with $g(y^*) = \lambda y^*$, $g(z^*) = \mu z^*$, $\lambda > 0 > \mu$.

5.3. Prove the following version of Lebesque's tiling Theorem [42]: For every n, there is an $\varepsilon > 0$ such that, whenever S^n is covered by a finite number of closed sets of diameter at most ε, there is a point of S^n lying in at least $n+1$ of the sets.

5.4. For $n \geq 2$ and $\varepsilon > 0$, construct a covering of S^n by finitely many closed sets of diameter at most ε so that no point lies in more than $n+1$ of the sets. (Hint: think of bricks for $n = 2$ and proceed inductively.)

5.5. Show that there is no triangulation of S^3 into more than one regular tetrahedron. (Hint: find the dihedral angle between adjacent faces of a tetrahedron.)

5.6. Devise an admissible labelling of the vertices of a triangulation, based on a continuous function f, so that the factor n in the inequality of 3.7 can be eliminated.

5.7. Prove Sperner's lemma directly from Brouwer's theorem. (Hint: if each vertex of the triangulation is assigned an image in S^n and each simplex of the triangulation is mapped linearly, the result is a continuous function. If a vertex labelled i is mapped to v^i fixed points do not necessarily lie in completely-labelled simplices--modify this rule to obtain the desired function.)

5.8. Let D^n denote the unit cube $\{x \in R^n_+ \mid x \leq u\}$ and let $f: D^n \to D^n$ be continuous. Define the following sets:

$$C'_0 = \{x \in D^n \mid f_j(x) > x_j \text{ or } x_j = 0 \text{ for all } j \in N\} \text{ and}$$

$$C_i' = \{x \in D^n \mid i = \min\{j \mid f_j(x) \leq x_j > 0 \text{ and } x_j \text{ is maximum satisfying}$$
$$\text{this condition}\}\} \text{ for each } i \in N.$$

(i) Show that the C_j', $j \in N_0$, partition D^n. Let $C_j = \overline{C_j'}$ for each $j \in N_0$. Show that a point in $\bigcap_{j \in N_0} C_j$ is a fixed point of f.

(ii) State and prove a lemma for D^n analogous to the K-K-M lemma for S^n. Use your lemma to prove the existence of a fixed point of f. (The sets in your lemma need not be based on a function f, but the properties they must satisfy can be motivated by the C_j's.) (Hint: Construct a homeomorphism h of D^n onto an n-simplex S so that each face of D^n is mapped into a subset of a face of S. Then apply the K-K-M lemma to S.)

(iii) State and prove a result analogous to 3.7 for D^n. (This exercise, and its continuations in III.6.4, IV.7.5, are related to the "cubical algorithms" of Allgower [1] and Jeppsen [31].)

5.9. Prove the following extension of Brouwer's Theorem: Let $T^n = \{x \in R^{n+1} \mid v^T x = 1\}$. Let $f: S^n \to T^n$ be continuous, and let $f_i(x) \geq 0$ whenever $x \in S_i^n$. Then f has a fixed point $x^* \in S^n$.

CHAPTER II: SOME APPLICATIONS OF BROUWER'S THEOREM

We will return later to the problem of constructing algorithms to locate completely-labelled simplices. Here we give some examples of the use of Brouwer's theorem in problems of interest. Of course, the existence result is less important to us than the construction of a function whose fixed points solve the given problem. We will also see some limitations of Brouwer's theorem that may suggest extensions.

If the theorem is applicable, finding a function with the required properties is usually straightforward. For any point that is not a solution, one generally has a good idea of how it should be modified. While an iterative algorithm based on successive modifications may oscillate and fail to converge, defining a function based on this modification and finding a fixed point generally solve the problem without difficulty.

II.1 A Model of an Exchange Economy. We will consider a very simplified model; the assumptions we make here will be partly relaxed in Chapter VI. A good discussion of the economic consequences of our assumptions can be found in Arrow and Hahn [2].

It is possible at this stage to include production, but only with the following very restrictive assumption. Faced with a set of prices, each producer picks a plan to maximize his profit; the required assumption would stipulate that his inputs and outputs are then continuous functions of the prices. We will be able to incorporate the possibility of production without this assumption in Chapter VI; for now we ignore production.

Suppose there are $n+1$ commodities, indexed by N_0. In the general model each commodity is a particular good or service at a particular location and date, and it is assumed that all possible futures markets exist. In this case economic agents make once-and-for-all decisions of all inputs and outputs for all time, and one must assume that they have perfect information. However, it is easier to visualize a bartering game played just once.

There are m economic agents, or consumers. Consumer i has an initial endowment of the commodities, forming a vector $w^i \in R_+^{n+1}$. He also has a preference relation $\geq_i$ ($x \geq_i z$ if consumer i prefers or is indifferent to x compared to z).

A price system is a nonzero vector $p \in R_+^{n+1}$, where (if $p_j > 0$) p_i/p_j units of commodity j are necessary to purchase one unit of commodity i. Given such a price system p, the budget set of consumer i is $B^i(p) = \{x \in R_+^{n+1} | p^T(x-w^i) \leq 0\}$. Consumer i will choose for consumption a commodity bundle $d^i(p)$ maximal with respect to $\geq_i$ in $B^i(p)$. It seems reasonable that $d^i(p)$ will exist and plausible that it will be unique if $p > 0$, i.e., all prices are positive (see exercise 4.1). We make the simplifying assumption that $d^i(p)$ is in fact defined and unique for all non-zero $p \geq 0$. (This may be acceptable if consumer i becomes satiated with each commodity.) It is clear that $d^i(\lambda p) = d^i(p)$ for say $\lambda > 0$. We therefore normalize the prices to add to one; in other words, we restrict p to S^n. Then $d^i: S^n \to R_+^{n+1}$ is a function, and we assume that d^i is continuous.

We further suppose that consumer i spends all his income, i.e., that $p^T d^i(p) = p^T w^i$ for all $p \in S^n$. This assumption is fairly innocuous, though it does conflict somewhat with the satiation referred to above. Then if $d(p) = \sum_{i=1}^m d^i(p)$ is the aggregate demand and $w = \sum_{i=1}^m w^i$ is the total initial endowment, we obtain $p^T d(p) = p^T w$. Let $g(p) = d(p) - w$ be the excess demand vector. Then $p^T g(p) = 0$, i.e., the value of the excess demand is zero, for all $p \in S^n$. This equality is known as Walras' Law, after the economist who first formulated this model of an economy. Also, since each d^i was assumed continuous, g is a continuous function from S^n to R^{n+1}.

A price system would be of little use unless it brought into harmony the disparate actions of all the consumers. Clearly, in order for all desired trades to be feasible, it is necessary and sufficient that $g(p)$ be non-positive.

Thus it is natural to state

__1.1 Definition.__ $p^* \in S^n$ is an equilibrium price vector if $g(p^*) \leq 0$.

Note that if Walras' Law holds and p^* is an equilibrium price vector, then $g_i(p^*) \leq 0$ for all $i \in N_0$ with equality if $p_i^* > 0$. Thus all markets are in balance except possibly those of free goods, which can be in excess supply.

We now show that under our assumptions an equilibrium price vector p^* always exists.

1.2 Theorem. Let $g: S^n \to R^{n+1}$ be continuous and satisfy $p^T g(p) = 0$ for all $p \in S^n$. Then there is a $p^* \in S^n$ with $g(p^*) \leq 0$.

Proof. We define a continuous function whose fixed points are equilibrium price vectors. To construct such a function, consider the following adjustment scheme for price vectors that are not equilibria: increase p_i if $g_i(p)$ is positive (commodity i is in excess demand) and decrease p_i if $g_i(p)$ is negative (commodity i is in excess supply). This classical "tâtonnement" may be globally unstable; but we will use it as the basis of our function. Consider the function h taking p into $p + \lambda g(p)$, where $\lambda > 0$. Unfortunately, h does not take S^n into itself and must be modified. First define $h^+: S^n \to R^{n+1}_+$ by $h^+(p) = (h_0^+(p), \ldots, h_n^+(p))^T$, with $h_i^+(p) = \max\{0, h_i(p)\}$. (We use this notation for the positive part of a vector throughout the present chapter.) For $p \in S^n$, $h^+(p)$ is nonnegative, but its coordinates may not sum to one. Finally, define $f: S^n \to S^n$ by setting $f(p) = h^+(p)/v^T h^+(p)$.

We must show that f is well-defined and continuous. Note that $v^T h^+(p) \geq 0$, with equality only if $h^+(p) = 0$. But then $h(p) \leq 0$ and $0 \geq p^T h(p) = p^T p + p^T g(p) = p^T p$ by Walras' Law. Since $p^T p > 0$ for all $p \in S^n$ we have a contradiction, establishing that $v^T h^+(p) > 0$ for all $p \in S^n$. Since g is continuous, so are h, h^+ and $v^T h^+$ (considered as a function from S^n to R). Hence $v^T h^+$, being continuous and positive on the compact set S^n, is bounded away from zero. We conclude that f is well-defined and continuous.

Brouwer's theorem guarantees a fixed point p^* of f. Thus $h^+(p^*) = \mu p^*$ for some $\mu > 0$. Let $I = \{i | p_i^* > 0\}$. We then have

(a) for all $i \in I$, $\mu p_i^* = h_i^+(p^*) = h_i(p^*) = p_i^* + \lambda g_i(p^*)$; and

(b) for all $j \notin I$, $0 = \mu p_j^* = h_j^+(p^*) \geq h_j(p^*) = \lambda g_j(p^*)$.

If $i \in I$, $g_i(p^*) = \rho p_i^*$ with $\rho = (\mu - 1)/\lambda$. Then Walras' Law gives $0 = p^{*T} g(p^*) = \sum_I p_i^* g_i(p^*) = \sum_I p_i^* \rho p_i^* = \rho p^{*T} p^*$. But $p^{*T} p^* > 0$, so $\rho = 0$ and $g_i(p^*) = 0$ for all $i \in I$. Combining this equality with the conclusion of (b), we have $g(p^*) \leq 0$, as required. □

1.3 Remarks.

(a) Although the proof that the fixed points of f are equilibrium price vectors is a little complicated, the construction of f is straightforward.

(b) The function f used above is taken from Arrow and Hahn; in contrast, Scarf [57,58] uses the function f' with $f'(p) = (p + g^+(p))/(1 + v^T g^+(p))$. The fixed points of f' are also equilibria, but for computational purposes f is superior. Let p^* be an equilibrium price vector with p_i^* positive but very small. Let p be close to p^* with $p_i > p_i^*$ and $g_i(p) < 0$. Then $f'(p)$ depends on the negative value of $g_i(p)$ only through Walras' Law, which requires some $g_j(p)$ to be positive and hence $f_i'(p) < p_i$. If p_i is small, this effect can be minimal. As a consequence, p can be very close to $f'(p)$ without $g(p)$ being correspondingly small. On the other hand, under these circumstances one can expect $g_i(p)$ to be fairly close to zero, and hence $f_i(p)$ will depend directly on $g_i(p)$. Since the scale of $f(p) - p$ is immaterial, λ should be chosen small so that f is more sensitive to all coordinates of g. To illustrate these ideas, we solved an example of Scarf [57; problem 1] using a modification of the algorithm of Chapter VIII. Using f with $\lambda = 1$ required 120 function evaluations; with $\lambda = 10^{-5}$ only 95 were necessary. Using f', with or without a scale factor multiplying g^+, the number required was 138, and the final excess demand vector was much larger.

(c) We proved theorem 1.2 using Brouwer's theorem. Uzawa [70] showed that the implication can be reversed: given $f: S^n \to S^n$, define g by $g(p) = f(p) - p(p^T f(p)/p^T p)$. Then g satisfies the hypotheses of 1.2. It is easy to verify that if $g(p^*) \leq 0$ then $f(p^*) = p^*$.

(d) In Chapter VI we extend our model of the economy to include production and demand correspondences (point-to-set mappings) instead of functions. The limitations of the present model will help to motivate extensions of Brouwer's theorem in Chapter V.

II.2 Unconstrained Optimization

Just as in section 1 we had to make simplifying assumptions to allow the use of Brouwer's theorem, here we will have to ignore constraints and consider only continuously differentiable functions.

Suppose we are trying to minimize a convex, continuously differentiable function $\theta: R^n \to R$ over R^n. For x^* to minimize θ it is necessary and sufficient that $\nabla\theta(x^*) = 0$. Thus a natural function is f with $f(x) = x - \nabla\theta(x)$. We need not worry about step size or convergence--but unfortunately there seems to be no closed n-cell that f takes into itself.

Let $C = \text{lev}_\alpha \theta = \{x \in R^n | \theta(x) \leq \alpha\}$ be bounded and assume that there is a $c \in \text{int } C$. Then I.2.2 shows that C is a closed n-cell, but one cannot assert that $f(C) \subseteq C$. We will have to resort to a device that is useful for proofs but is not much help in computation. (Another approach is given in exercise 4.2.) The difficulty will be removed in Chapter VI.

Since C is compact, so is $f(C)$; embed $C \cup f(C)$ in a large simplex S. We have $f: C \to S$ and we want to extend f to, say, $\bar{f}: S \to S$ without creating new fixed points. Choose $r: S \to C$ to satisfy:

(i) r is continuous;

(ii) on C, r is the identity; and

(iii) if $x \notin C$ and $\lambda \in (0,1]$, $\lambda x + (1-\lambda)r(x) \notin C$.

A function satisfying (i) and (ii) is called a retraction of S onto C; we will give two examples of functions r satisfying (i)-(iii) below.

Now let $\bar{f}: S \to S$ be $f \circ r$. $\bar{f}$ is continuous and hence has a fixed point x^*. If $x^* \in C$, $r(x^*) = x^*$, $f(x^*) = x^*$, and $\nabla\theta(x^*) = 0$. We now show that $x^* \notin C$ leads to a contradiction. Let $z = r(x^*)$. Then $x^* = z - \nabla\theta(z)$. But for sufficiently small λ, $z(\lambda) = \lambda x^* + (1-\lambda)z = z - \lambda\nabla\theta(z)$ has $\theta(z(\lambda)) < \theta(z)$; hence $z(\lambda) \in C$, contradicting condition (iii) of r. Thus any fixed point of $\bar{f}$ is a fixed point of f and a minimizer of θ.

We have proved that a continuous function achieves its minimum on a compact set --a result provable by far simpler means. The result is even less impressive when we note that finding the minimizer will be difficult even if we have good algorithms for finding fixed points because of the retraction r. Even in the simplest case evaluation of r requires a line search. Again we are motivated to extend Brouwer's theorem, to encompass both functions on R^n and point-to-set mappings (if θ is not continuously differentiable).

Two choices for r are r^1 and r^2, defined below:

(a) let $r^1(x) \in C$ satisfy $\|r^1(x) - x\|_2 \leq \|z-x\|_2$ for each $z \in C$;

(b) let $r^2(x) = \lambda x + (1-\lambda)c$, where $\lambda = \max\{\mu \leq 1 | \mu x + (1-\mu)c \in C\}$.

II.3 The Nonlinear Complementarity Problem

Let $g: R^n_+ \to R^n$ be continuous. The nonlinear complementarity problem (NLCP) associated with g is to find $x^* \geq 0$ with $g(x^*) \geq 0$ and $x^{*T}g(x^*) = 0$. If g is affine $(g(x) = Ax + b)$, this is the well-known linear complementarity problem--see Lemke's survey paper [44]. The NLCP was first studied by Cottle [7].

One application of the NLCP is to nonlinear programming. Consider the problem: $\text{minimize}\{\theta(z) | g_i(z) \leq 0,\ i = 1,2,\ldots,m,\ z \in R^k_+\}$. Let $n = k + m$ and $x = (z,w)$, where $w \in R^m$ is a vector of Kuhn-Tucker multipliers. When θ and all the g_i's are continuously differentiable, finding a Kuhn-Tucker point is equivalent to solving the NLCP associated with h, where $h(z,w) = (\nabla\theta(z) + w^T\nabla g(z), -g(z))$.

The following result was proved by Moré [50]:

<u>3.1 Theorem</u>. Suppose there are $w \in R^n_+$ and $\rho > \|w\|_\infty$ such that, for all $x \geq 0$ with $\|x\|_\infty = \rho$, $\max_{i\in N} (x_i-w_i)g_i(x) > 0$. Then the NLCP associated with g has a solution in $C_\rho = \{x \in R^n_+ | \|x\|_\infty \leq \rho\}$.

<u>Proof</u>. First we define a function $h: R^n_+ \to R^n_+$ by $h(x) = (x - g(x))^+$. Then x^* is a fixed point of h iff it is a solution to the NLCP.

Since h is continuous, $C' = C_\rho \cup h(C_\rho)$ is compact; choose $\tau > 0$ so that $C' \subseteq C_\tau$. Note that C_τ is a closed n-cell. Define the retraction $r: C_\tau \to C_\rho$ as follows. If $x \in C_\rho$, $r(x) = x$; otherwise, let $\lambda = \lambda(x) = \min\{(\rho-w_i)/(x_i-w_i) | x_i > \rho\}$ and $r(x) = \lambda x + (1-\lambda)w$. Then r is the radial retraction onto C_ρ with center w; clearly λ and hence r are continuous functions of x. Now let $f(x) = h(r(x))$; then $f: C_\tau \to C_\tau$ is continuous.

Brouwer's theorem yields a fixed point x^* of f. Suppose that $x^* \notin C_\rho$. Let $z^* = r(x^*)$ and note that $z^* \geq 0$, $\|z^*\|_\infty = \rho$. Choose i so that $(z^*_i-w_i)g_i(z^*) > 0$. Then either $z^*_i > w_i$, in which case $x^*_i \geq z^*_i > w_i \geq 0$ and $x^*_i \geq z^*_i > z^*_i - g_i(z^*)$; or $z^*_i < w_i$ and $x^*_i \leq z^*_i < z^*_i - g_i(z^*) > 0$. In either case

$x_i^* \neq (z_i^* - g_i(z^*))^+ = f_i(x^*)$. Hence $x^* \in C_\rho$, $x^* = h(x^*)$ and x^* solves the NLCP associated with g. □

We chose the existence result above because it is relatively strong and also simple to prove using Brouwer's theorem. For other results and applications see Karamardian [33], Eaves [11], and Fisher and Gould [59].

II.4 Exercises

4.1. Let the binary relation $\succeq$, defined on R_+^{n+1}, satisfy: for all $x,y,z \in R_+^{n+1}$, (i) $x \succeq y$ or $y \succeq x$; (ii) $x \succeq y$ and $y \succeq z$ imply $x \succeq z$; and (iii) $U_x = \{x' \in R_+^{n+1} | x' \succeq x\}$ and $L_x = \{x' \in R^{n+1} | x \succeq x'\}$ are closed, and U_x is convex.

Let $w \in R_+^{n+1}$ be fixed, and let T^n denote $\{p \in S^n | p > 0\}$. For all $p \in T^n$, let $B(p) = \{x \in R_+^{n+1} | p^T(x-w) \leq 0\}$ and $D(p) = \{x \in B(p) | x \succeq y$ for all $y \in B(p)\}$. Show that $D(p)$ is nonempty and convex for all $p \in T^n$.

Now suppose further that for all $x,y \in R_+^{n+1}$, $x \neq y$, either $x \succeq (x+y)/2$ or $y \succeq (x+y)/2$ is false. Show that $D(p)$ is a singleton $\{d(p)\}$ and that $d: T^n \to R_+^{n+1}$ is continuous.

4.2. Let $\theta: R^n \to R$ be convex and continuously differentiable, with $C = \text{lev}_\alpha \theta = \{x \in R^n | \theta(x) \leq \alpha\}$ bounded and $c \in \text{int } C$. Show that there is some $\lambda > 0$ so that, if $f(x) = x - \lambda\nabla\theta(x)$, f takes C into itself. (See also I.5.9.)

CHAPTER III: TRIANGULATIONS

We now present formally the discussion of triangulations, including a proof of the properties used in I.3.4. We also introduce some particular triangulations that are used in Chapter IV for the development of algorithms for finding completely-labelled simplices.

Section 1 discusses affine independence and simplices and Section 2 introduces triangulations and derives some of their properties. In Section 3 we define some particular triangulations and in Section 4 describe their pivot rules. Section 5 briefly presents Scarf's primitive sets [57,58].

III.1 Affine Independence and Simplices. Given points $x^0,\ldots,x^j$ in R^m, we can form the combination $x = \sum_{i=0}^{j} \lambda_i x^i$, with all $\lambda_i \in R$. Then x is a linear combination of the x^i's. Also familiar are nonnegative combinations, where each $\lambda_i \in R_+$, and convex combinations, where each $\lambda_i \in R_+$ and $\sum_{i=0}^{j} \lambda_i = 1$. We say x is an affine combination of the x^i's if $\sum_{i=0}^{j} \lambda_i = 1$, but the λ_i's are not necessarily nonnegative.

1.1 Definition. If $S \subseteq R^m$, the affine hull of S, denoted $\text{aff}(S)$, is the set of all finite affine combinations of members of S. S is an affine subspace if $\text{aff}(S) = S$. The dimension of $S \neq \emptyset$ is the dimension of the linear subspace parallel to $\text{aff}(S)$, i.e., $\text{aff}(S) - \{s\}$ for any $s \in S$. (If S is empty, $\dim(S) = -1$.)

All relative concepts of topology are relative to the affine hull. Thus the relative interior of S, rel int S, is $\{x \in S \mid \exists\, \varepsilon > 0 \text{ s.t. } B(x,\varepsilon) \cap \text{aff}(S) \subseteq S\}$. In particular, we mean the relative boundary of a set S whenever we refer to its boundary and denote it by ∂S. Thus $\partial S = \overline{S} \cap \overline{(\text{aff}(S) \sim S)}$. This notation is consistent with the definition of ∂S^n in I.2.1.

Just as linearly independent points have the property that none is a linear combination of the rest, affinely independent points are defined to have the analogous property. An alternative statement is

1.2 Definition. $y^0,\ldots,y^j$ in R^m are affinely independent if $\sum_{i=0}^{j} \lambda_i y^i = 0$

and $\sum_{i=0}^{j} \lambda_i = 0$ imply $\lambda_i = 0$ for all i. One may easily verify the following result.

1.3 Lemma. The following are equivalent for $y^0,\ldots,y^j \in R^m$:

(a) $y^0,\ldots,y^j$ are affinely independent;

(b) the matrix $Y = \begin{bmatrix} 1 & \ldots & 1 \\ y^0 & \ldots & y^j \end{bmatrix}$ has rank $j+1$;

(c) the vectors $z^i = y^i - y^{i-1}$, $1 \leq i \leq j$, are linearly independent; and

(d) $\dim(\{y^0,\ldots,y^j\}) = j$. □

1.4 Example. $v^0,\ldots,v^n$ are affinely independent in R^{n+1}, and their affine hull is $\{x \in R^{n+1} | v^T x = 1\}$.

We can now define simplices.

1.5 Definition. If $y^0,\ldots,y^j$ in R^m are affinely independent, then the relative interior of their convex hull, i.e., $\sigma = \{\sum_{i=0}^{j} \lambda_i y^i | \lambda_i > 0$ for all i, $\sum_{i=0}^{j} \lambda_i = 1\}$ is a j-simplex. The vertices of σ are $y^0,\ldots,y^j$. We write $\sigma = \langle y^0,\ldots,y^j \rangle$. Note that σ is relatively open. Denote by $\bar{\sigma} = [y^0,\ldots,y^j]$ the closure of σ -- $\bar{\sigma}$ is called a closed j-simplex.

Note: From 1.3 no simplex in R^m has dimension more than m.

We will not distinguish between a 0-simplex $[y]$ or $\langle y \rangle$ and its vertex y.

1.6 Definition. A simplex τ is a face of a simplex σ if all the vertices of τ are vertices of σ. In particular, σ is an improper face of itself. If $\dim \tau = \dim \sigma - 1$, τ is called a facet of σ. If y is the vertex of σ which is not a vertex of τ, τ is the facet of σ opposite y.

1.7 Example (m = 2).

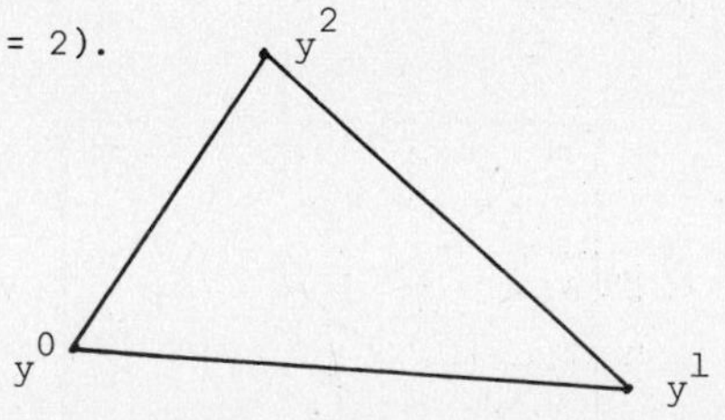

$\sigma = \langle y^0, y^1, y^2\rangle$ is the interior of the triangle; $\tau = \langle y^0, y^1\rangle$ is the open segment from y^0 to y^1; $\tau' = \langle y^0\rangle = [y^0]$ is just $\{y^0\}$. τ' and τ are faces of σ, and τ' is a facet of τ.

The following important result is easy to prove:

1.8 Lemma. $\bar{\sigma}$ is partitioned by all the faces of σ. □

III.2 Triangulations. Let C be a convex set in R^m. Let $A = \mathrm{aff}(C)$ have dimension $n \leq m$. (Note that to triangulate S^n we have to consider the case $n < m$.)

2.1 Definition. G is a triangulation of C if

(i) G is a collection of n-simplices;

(ii) the faces of all the simplices in G partition C; and

(iii) each $x \in C$ has a neighborhood meeting only a finite number of simplices of G.

Condition (iii) stipulates that G is locally finite. Denote by G^j the collection of j-simplices that are faces of simplices of G. Thus $G^n = G$. Call members of G^0 vertices of G. Denote by G^+ all faces of simplices of G, i.e., $\bigcup_{j \in N_0} G^j$.

2.2 Examples.

(a) $G = \{\langle y^0,y^1,y^2\rangle, \langle y^0,y^2,y^3\rangle, \langle y^0,y^3,y^4\rangle, \langle y^0,y^4,y^1\rangle\}$ triangulates the square:

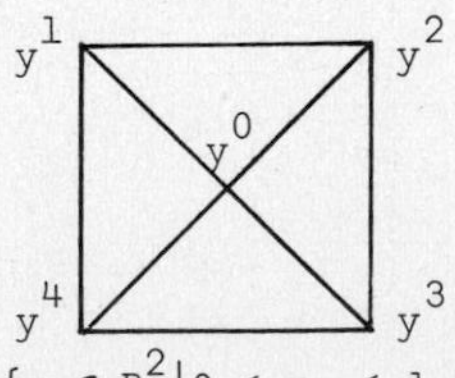

(b) Let $C = \{x \in R^2 | 0 \leq x_1 \leq 1,\ 0 < x_2 \leq 1\}$. C can be triangulated as follows:

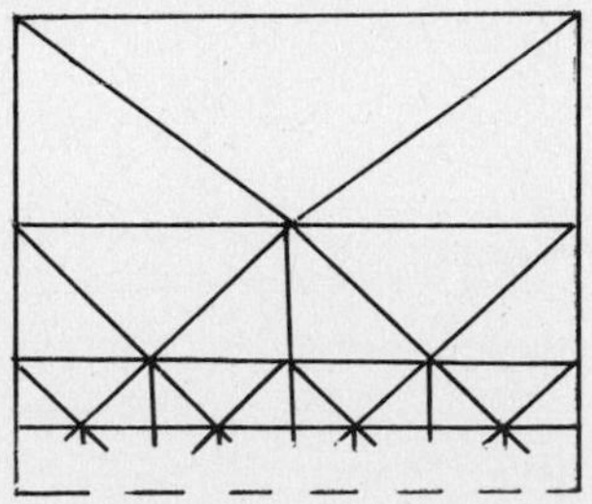

This example shows that C need not be closed and, even if C is bounded, G need not be a finite collection.

We summarize the most important properties of triangulations below. Note that (a) and (b) give the alternative characterization by closed simplices, and (d) and (e) are I.3.4(a) and (c).

<u>2.3 Theorem</u>. Suppose G is a triangulation of C with $\dim C = n$. Then

(a) $\bigcup_{\sigma \in G} \bar{\sigma} = C$.

(b) If $\sigma_1, \sigma_2 \in G$ and $\bar{\sigma}_1 \cap \bar{\sigma}_2 \neq \emptyset$, then $\bar{\sigma}_1 \cap \bar{\sigma}_2$ is the closure of a common face τ of σ_1 and σ_2.

(c) If $D \subseteq C$ is compact, D meets only finitely many simplices of G.

(d) If $\tau \in G^{n-1}$, then either

(i) $\tau \subseteq \partial C$ and τ is a facet of just one simplex of G; or

(ii) $\tau \not\subseteq \partial C$ and τ is a facet of exactly two simplices of G.

(e) Let $D \subseteq \partial C$, $\dim D = n-1$, and $D = C \cap \text{aff}(D)$. Then $G' = \{\tau | \tau \subseteq D, \tau \in G^{n-1}\}$ triangulates D.

<u>Proof</u>.

(a) By definition C is the union of all faces of all simplices of G. By 1.8 $\bar{\sigma}$ is the union of all faces of σ. Thus part (a) follows.

(b) Let $x \in \bar{\sigma}_1 \cap \bar{\sigma}_2$. Then by 1.8 x lies in a unique face τ_1 of σ_1 and a unique face τ_2 of σ_2. Since G is a triangulation, $\tau_1 = \tau_2$, i.e., each point of $\bar{\sigma}_1 \cap \bar{\sigma}_2$ lies in a common face of σ_1 and σ_2. Let $w^0, \ldots, w^j$ be a list of all vertices of such common faces. Then each w^i is a common vertex of σ_1 and σ_2. Each $x \in \bar{\sigma}_1 \cap \bar{\sigma}_2$ lies in $[w^0, \ldots, w^j]$ and clearly $[w^0, \ldots, w^j] \subseteq \bar{\sigma}_1 \cap \bar{\sigma}_2$ since the latter is convex. Hence $\bar{\sigma}_1 \cap \bar{\sigma}_2 = [w^0, \ldots, w^j]$, a common face of σ_1 and σ_2.

(c) By 2.1(iii), each $x \in D$ has a neighborhood meeting only finitely many simplices of G. Clearly, these neighborhoods cover D; hence there is a finite subcover. Thus only a finite number of simplices of G meet D.

(d) Since $\dim C = n > n-1 = \text{aff}(\tau)$, there is a point $w \in C \sim \text{aff}(\tau)$. Let x be the barycenter of τ; i.e., if $\tau = \langle y^1, y^2, \ldots, y^n \rangle$, $x = \sum_{i \in N} y^i/n$. For

$\ell = 1,2,\ldots$ let $x^\ell = w/\ell + (1-1/\ell)x$. Then $x^\ell \in C$ for all ℓ, $x^\ell \to x$, and no x^ℓ lies in $\mathrm{aff}(\tau)$. There is a neighborhood M of x meeting only finitely many simplices of G but infinitely many x^ℓ; hence there is a simplex σ in G^+ containing infinitely many x^ℓ. Since $x^\ell \to x$, $x \in \bar{\sigma}$ and $x \in \bar{\tau}$; thus by (b) x lies in the closure of a common face of σ and τ. But x lies in τ itself, so τ is a face of σ. Since $\sigma \neq \tau$, σ is an n-simplex of G. Thus each $\tau \in G^{n-1}$ is a facet of at least one $\sigma \in G$.

(i) Let $\tau \subseteq \partial C$ and assume τ is a facet of σ_1 and σ_2, $\sigma_1 \neq \sigma_2$. We obtain a contradiction. Let $\sigma_1 = \langle z^0, y^1, \ldots, y^n \rangle$ and $\sigma_2 = \langle w^0, y^1, \ldots, y^n \rangle$. Since σ_1 is an n-simplex, $\mathrm{aff}(\sigma_1) = \mathrm{aff}(C)$ contains w^0; hence we have $w^0 = \rho z^0 + \sum_{i \in N} \lambda_i y^i$ with $\rho + \sum_{i \in N} \lambda_i = 1$.

Distinguish three cases:

<u>Case 1: $\rho > 0$</u>. Then consider $c = (1-\varepsilon)x + \varepsilon w^0 = \varepsilon w^0 + \sum_{i \in N} ((1-\varepsilon)/n) y^i = \rho\varepsilon z^0 + \sum_{i \in N} (\varepsilon\lambda_i + (1-\varepsilon)/n) y^i$. For any $0 < \varepsilon < 1$, $c \in \sigma_2$, as shown by the first expression. But for sufficiently small ε each coefficient in the second expression is positive and $c \in \sigma_1$, a contradiction.

<u>Case 2: $\rho = 0$</u>. Then w^0 is an affine combination of $y^1, \ldots, y^n$ and σ_2 is not an n-simplex.

<u>Case 3: $\rho < 0$</u>. Every point c in $\mathrm{aff}(\sigma_2) = \mathrm{aff}(C)$ can be expressed uniquely as $c = \pi w^0 + \sum_{i \in N} \mu_i y^i$, $\pi + \sum_{i \in N} \mu_i = 1$. By substituting we find $c = \rho\pi z^0 + \sum_{i \in N} (\mu_i + \pi\lambda_i) y^i$ also. If c is close to x, then π is close to zero and μ_i is close to $1/n$, for $i \in N$. More precisely, x has a neighborhood M in $\mathrm{aff}(C)$ so that if $c \in M$, $\mu_i \geq 1/2n$ and $|\pi| \leq 1/3n\lambda$, where $\lambda = \max_{i \in N} |\lambda_i|$. But then any point of M lies in σ_2 (if $\pi > 0$), τ (if $\pi = 0$) or σ_1 (if $\pi < 0$), so that $M \subseteq C$. This contradicts $x \in \tau \subseteq \partial C$.

(ii) $\tau \not\subseteq \partial C$. Then there is an $x' \in \tau$ with a neighborhood (in $\mathrm{aff}(C)$) contained in C. From this neighborhood we can pick two points w^1 and w^2 with $(w^1+w^2)/2 = x'$ and $w^1, w^2 \notin \mathrm{aff}(\tau)$. Thus w^1 and w^2 are on opposite sides of τ. By the construction above we can find two simplices σ_1 and σ_2 with τ as a facet, and it is easy to show that $\sigma_1 \neq \sigma_2$. The arguments of case 1 show that there can be no third simplex.

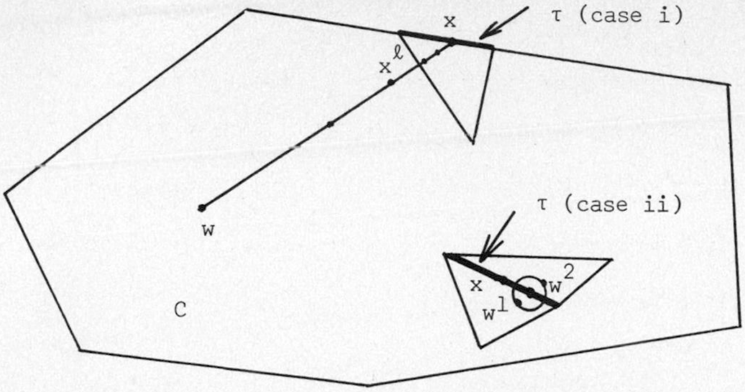

(e) G' consists of (n-1)-simplices in G^{n-1}. Clearly, all faces of simplices of G' are disjoint, and the local finiteness condition holds for G' since it holds for G. Thus we need only prove that the simplices of G' and their faces cover D. Pick any $x \in D$; since $x \in C$, x lies in some $\rho \in G^j$. Because x lies in ∂C, it has no neighborhood lying in C so that ρ cannot be a n-simplex. If any vertex of ρ did not lie in $\text{aff}(D)$, then there would have to be vertices of ρ on either side of $\text{aff}(D)$. This contradicts $D \subseteq \partial C$. Thus each vertex of ρ lies in $\text{aff}(D)$ and hence in D; we deduce that $\rho \subseteq D$. If ρ is a (n-1)-simplex, we are through; otherwise, use the technique of (d) to find a higher-dimensional simplex in $\text{aff}(D)$ with ρ as a face. Eventually a (n-1)-simplex in G' is found; hence ρ is a face of a simplex in G'. □

III.3 Some Special Triangulations of R^n and S^n. For consistency of notation we sometimes use names for triangulations that have been applied elsewhere in the literature.

3.1 The Triangulation K_1 of R^n. This triangulation is often called K or "Kuhn's triangulation." Kuhn [37] credits it to Tucker [43, page 140], but it appears earlier in Freudenthal [23]. Freudenthal constructed K_1 to answer a question of Brouwer: do there exist regular triangulations of a simplex of arbitrary mesh whose simplices do not become (in a precise sense) "arbitrarily long and skinny"? Whether Brouwer was considering algorithms to compute fixed points is a fascinating question.

Note that our K_1 is not the same as Eaves' K_1 in [14].

Let $K_1^0 = Z^n = \{y \in R^n | y^i \in Z$ for each $i \in N\}$. If $y^0 \in K_1^0$ and π is a permutation of N, then denote by $k_1(y^0,\pi)$ the n-simplex $\langle y^0,\ldots,y^n \rangle$ where $y^i = y^{i-1} + u^{\pi(i)}$ for each $i \in N$. (1.3(c) shows that $k_1(y^0,\pi)$ is an n-simplex.) Finally, let K_1 be the collection of all such $k_1(y^0,\pi)$. Whenever we set $k_1(y^0,\pi) = \langle y^0,\ldots,y^n \rangle$, we suppose the y^i's ordered as above, and similarly for the other triangulations below.

3.2 Lemma. K_1 is a triangulation of R^n.

Proof. Clearly, K_1 is a collection of n-simplices and any point of R^n has a neighborhood meeting only finitely many simplices of K_1. We need only prove that the simplices of K_1^+ partition R^n. Let x be an arbitrary point of R^n. For each $i \in N$, let $y_i^0 = \lfloor x_i \rfloor$ where, for any real λ, $\lfloor \lambda \rfloor$ denotes the greatest integer not exceeding λ. Then $y^0 \in K_1^0$. Let $z = y^0 - x$; we have $0 \leq z \leq u$.

Let π be a permutation of N such that $1 \geq z_{\pi(1)} \geq \cdots \geq z_{\pi(n)} \geq 0$. Denote the terms above by $\alpha_0 = 1, \alpha_1,\ldots,\alpha_n, \alpha_{n+1} = 0$. For $i \in N_0$ let $\beta_i = \alpha_i - \alpha_{i+1}$. Then each $\beta_i \geq 0$ and $\sum_{i\in N_0} \beta_i = \alpha_0 - \alpha_{n+1} = 1$.

Let $\sigma = \langle y^0,\ldots,y^n \rangle = k_1(y^0,\pi)$ and consider $\sum_{i\in N_0} \beta_i y^i$. We have

$$\sum_{i\in N_0} \beta_i y^i = y^0 + \sum_{i\in N} \beta_i (y^i - y^0) = y^0 + \sum_{i\in N} \beta_i (\sum_{j=1}^{i} (y^j - y^{j-1}))$$

$$= y^0 + \sum_{j\in N} (y^j - y^{j-1})(\sum_{i=j}^{n} \beta_i) = y^0 + \sum_{j\in N} u^{\pi(j)} \alpha_j$$

$$= y^0 + \sum_{j\in N} u^{\pi(j)} z_{\pi(j)} = y^0 + z = x.$$

Thus $x \in \bar{\sigma}$. Using 1.8, we deduce that the simplices of K_1^+ cover R^n.

We must now show that all these faces are disjoint. Again let $x \in R^n$ be arbitrary. We may assume that $x \in \bar{\sigma}$ for $\sigma = \langle y^0,\ldots,y^n \rangle = k_1(y^0,\pi) \in K_1$, say $x = \sum_{i\in N_0} \beta_i y^i$, $\beta_i \geq 0$ for $i \in N_0$ and $\sum_{i\in N_0} \beta_i = 1$. Then x lies in a face of σ whose vertices are those y^ℓ for $\ell \in L = \{\ell \in N_0 | \beta_\ell > 0\}$. We show below how each y^ℓ, $\ell \in L$, can be generated from x independently of y^0 and π. Thus these same vertices are found for any simplex of K_1 whose closure contains x. Hence x

lies in just one face of a simplex of K_1, and the argument is complete.

For any $z \in \bar{\sigma}$, we have $z = \sum_{i\in N_0} \beta_i(z)y^i$, say, with $\beta_i(z) \geq 0$ for $i \in N_0$ and $\sum_{i\in N_0} \beta_i(z) = 1$. We then form $\alpha_i(z) = \sum_{j=i}^{n} \beta_j(z)$ for $i - 0,1,\ldots,n+1$ (conventionally $\alpha_{n+1}(z) = 0$). By the argument above in reverse, we have $1 = \alpha_0(z) \geq \ldots \geq \alpha_{n+1}(z) = 0$ with $\alpha_i(z) = z_{\pi(i)} - y_{\pi(i)}$ for $i \in N$. This is true for x and also for each y^ℓ, $\ell \in L$. Then we find that for $\ell \in L$, $(\alpha_0(y^\ell)\ldots\alpha_{n+1}(y^\ell))$ has the form $(1,1,\ldots,1,0,0,\ldots,0)$ with the last 1 in position ℓ. We can now, with our knowledge of the "α-vectors" of x and each y^ℓ, construct all the y^ℓ, $\ell \in L$, directly from x.

For each $\gamma \in (0,1]$, let $y(\gamma)$ (one of the y^ℓ's) be defined by

$$y^i(\gamma) = \begin{cases} \lfloor x_i \rfloor + 1 & \text{if } x_i - \lfloor x_i \rfloor \geq \gamma \\ \lfloor x_i \rfloor & \text{if } x_i - \lfloor x_i \rfloor < \gamma. \end{cases}$$

It is clear that each y^ℓ is of this form; also there will be one distinct $y(\gamma)$ for each strict inequality in $\alpha_0(x) \geq \ldots \geq \alpha_{n+1}(x)$, and thus $|L|$ in total.

Noting that the $y(\gamma)$'s can be determined from x alone without reference to y^0 or π, we have completed the proof of the lemma. □

<u>3.3 Example</u> $(n = 4)$. Let $x = (2\frac{1}{2}, 1\frac{1}{2}, -2\frac{1}{4}, \frac{1}{4})^T$. Let us find all simplices of K_1 whose closures contain x. We have $y = (2,1,-3,0)^T$ and $z = x-y = (\frac{1}{2}, \frac{1}{2}, \frac{3}{4}, \frac{1}{4})$. We can choose $\pi^1 = (3,1,2,4)$ or $\pi^2 = (3,2,1,4)$. In either case $\alpha = (1, \frac{3}{4}, \frac{1}{2}, \frac{1}{2}, \frac{1}{4}, 0)$ and $\beta = (\frac{1}{4}, \frac{1}{4}, 0, \frac{1}{4}, \frac{1}{4})$. The two possible simplices are $\langle y^0,\ldots,y^4\rangle = k_1(y,\pi^1)$ and $\langle z^0,\ldots,z^4\rangle = k_1(y,\pi^2)$ where $y^0 = z^0 = (2,1,-3,0)^T$, $y^1 = z^1 = (2,1,-2,0)^T$, $y^2 = (3,1,-2,0)^T$, $z^2 = (2,2,-2,0)^T$, $y^3 = z^3 = (3,2,-2,0)^T$ and $y^4 = z^4 = (3,2,-2,1)^T$. The reader can confirm that $x = \sum \beta_i y^i = \sum \beta_i z^i$. Note that x lies in the common face $\langle y^0,y^1,y^3,y^4\rangle$.

We can also proceed directly from x to determine the vertices of the face of K_1 containing x, as in the last part of the proof. For $0 < \gamma \leq \frac{1}{4}$, $y(\gamma) = (3,2,-2,1)^T = y^4$. For $\frac{1}{4} < \gamma \leq \frac{1}{2}$, $y(\gamma) = (3,2,-2,0)^T = y^3$. For $\frac{1}{2} < \gamma \leq \frac{3}{4}$, $y(\gamma) = (2,1,-2,0)^T = y^1$. Finally, for $\frac{3}{4} < \gamma \leq 1$, $y(\gamma) = (2,1,-3,0) = y^0$.

3.4. Note that the crucial part of the proof of 3.2 was obtaining the alternative description of $\overline{k_1(y,\pi)}$ as those x with

$$1 \geq x_{\pi(1)} - y_{\pi(1)} \geq \cdots \geq x_{\pi(n)} - y_{\pi(n)} \geq 0. \qquad (*)$$

It is easy to see that $k_1(y,\pi)$ consists of those x satisfying (*) with strict inequalities throughout. By requiring all inequalities but one to be strict and replacing the remaining inequality by an equality, we obtain the set of x lying in a particular facet of (*). For this reason we refer to (*) as a facetal description of $k_1(y,\pi)$. From (*) we notice that $x \in R^n$ lies in a simplex of K_1^+ of dimension less than n if and only if some x_i is integer or some $(x_i - x_j)$ is integer.

3.5 The Triangulation J_1 of R^n. A second triangulation of R^n is J_1, originally due to Tucker [43, page 140] and also found in Whitney [72, page 358]. See also [64].

Let $J_1^0 = Z^n$. Distinguish certain members of J_1^0 (to be called "central vertices") by denoting by J_1^{0c} the set $\{y \in J_1^0 | y_i$ is odd for each $i \in N\}$. If $y^0 \in J_1^{0c}$, π is a permutation of N, and $s \in R^n$ is a sign vector (each s_i is ± 1), then denote by $j_1(y^0,\pi,s)$ the n-simplex $\langle y^0,\ldots,y^n\rangle$ where $y^i = y^{i-1} + s_{\pi(i)}u^{\pi(i)}$ for each $i \in N$. Let J_1 be the collection of all such $j_1(y^0,\pi,s)$.

We leave it as an exercise to prove that J_1 is a triangulation. The proof follows that for K_1; the key string of inequalities analogous to (*) is here

$$1 \geq s_{\pi(1)}(x_{\pi(1)} - y_{\pi(1)}) \geq \cdots \geq s_{\pi(n)}(x_{\pi(n)} - y_{\pi(n)}) \geq 0. \qquad (**)$$

3.6 Obtaining triangulations of S^n. For any $C \subseteq R^n$ and $\delta \in R$, δC denotes $\{\delta c | c \in C\}$. For any family G of subsets of R^n and any $0 \neq \delta \in R$, let δG denote $\{\delta C | C \in G\}$; if G is a triangulation of D, δG is a triangulation of δD. In particular, we now have triangulations δK_1 and δJ_1 of R^n. Now recall that $\text{mesh}_p G = \sup_{C \in G} \text{diam}_p C$. Hence we easily obtain from the definitions that $\text{mesh}_2 \delta K_1 = \text{mesh}_2 \delta J_1 = \sqrt{n}\,|\delta|$ and $\text{mesh}_\infty \delta K_1 = \text{mesh}_\infty \delta J_1 = |\delta|$. Thus we have triangulations of R^n of arbitrary mesh. These generate triangulations of S^n as follows. Denote by C^n the set $\{x \in R^n | 1 \geq x_1 \geq \ldots \geq x_n \geq 0\}$ as in I.2.3. We first triangulate C^n and then use the method of I.2.3 to obtain triangulations of S^n.

Choose $0 < m \in Z$ and let $\tilde{K}_1$ ($\tilde{J}_1$) be the collection of n-simplices of K_1 (J_1) meeting $mC^n = \{x \in R^n | m \geq x_1 \geq \ldots \geq x_n \geq 0\}$. Let $x \in mC^n$ lie in $\sigma \in K_1$ ($\sigma \in J_1$). As in the proof of 3.2, each vertex of σ can be obtained from x by sliding each coordinate of x up or down to an adjacent integer. Hence each such vertex also lies in mC^n, and it is easy to deduce that $\tilde{K}_1$ and $\tilde{J}_1$ triangulate mC^n, and that $m^{-1}\tilde{K}_1$ and $m^{-1}\tilde{J}_1$ triangulate C^n. Finally, using the affine homeomorphism h of I.2.3, we obtain the triangulations $K_2(m)$ and $J_2(m)$ of S^n.

<u>3.7 Definition</u>. Let Q denote the $(n+1)\times n$ matrix $\begin{bmatrix} -1 & & 0 \\ +1 & \ddots & \\ & \ddots & -1 \\ 0 & & +1 \end{bmatrix}$ and q^j its jth column for $j \in N$. Let $K_2^0(m) = \{y \in S^n | my_i \in Z\}$ for each $i \in N_0$. If $y^0 \in K_2^0(m)$ and π is a permutation of N, let $\sigma = \langle y^0,\ldots,y^n\rangle$ where $y^i = y^{i-1} + m^{-1}q^{\pi(i)}$ for each $i \in N$. If $\sigma \subseteq S^n$, we write $\sigma = k_2(y^0,\pi)$ (m is implicit). Finally, let $K_2(m)$ be the collection of all such $k_2(y^0,\pi)$.

<u>3.8 Definition</u>. Let $J_2^0(m) = K_2^0(m)$ and $J_2^{0c}(m) = \{y \in J_2^0(m) | my_i$ is even, $1 \leq i < n$ and my_n is odd$\}$. If $y^0 \in J_2^{0c}(m)$, π is a permutation of N, and $s \in R^n$ is a sign vector, let $\sigma = \langle y^0,\ldots,y^n\rangle$ where $y^i = y^{i-1} + m^{-1}s_{\pi(i)}q^{\pi(i)}$ for each $i \in N$. If $\sigma \subseteq S^n$, we write $\sigma = j_2(y^0,\pi,s)$. Finally, let $J_2(m)$ be the collection of all such $j_2(y^0,\pi,s)$.

<u>3.9 Lemma</u>. $K_2(m)$ and $J_2(m)$ are triangulations of S^n. $\text{mesh}_\infty K_2(m) = m^{-1}$, $\text{mesh}_\infty J_2(m) = 2m^{-1}$, $\text{mesh}_2 K_2(m) = m^{-1}\sqrt{n+1}$ if n is odd, $m^{-1}\sqrt{n}$ if n is even, and $\text{mesh}_2 J_2(m) = m^{-1}\sqrt{4n-2}$. □

3.9 and 2.3(d), (e) establish the earlier properties I.3.4(a)-(c) used in the proof of Sperner's lemma. The proof of Brouwer's theorem is finally complete!

<u>3.10 Remarks</u>. For m a power of 2, $J_2(m)$ is due to Whitney [72, pp. 358-60].

Note that we have not mentioned the triangulation most familiar to topologists, that of iterated barycentric subdivision. The example below shows clearly how this triangulation progresses.

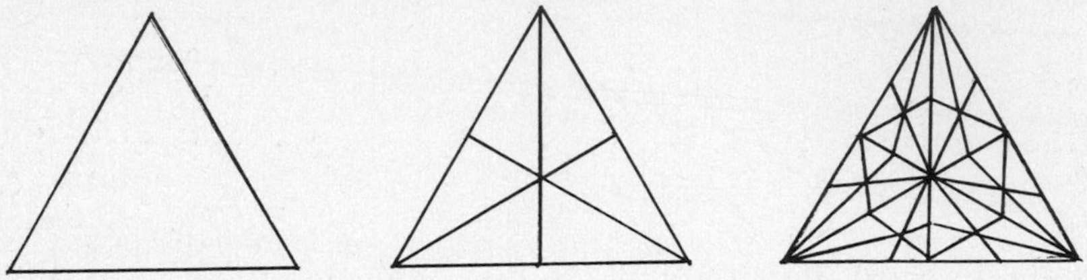

The reason for this seemingly glaring omission is that this triangulation exhibits the inefficient behavior of creating long, skinny simplices--an algorithm generates a large number of simplices ending in one of relatively large diameter, which gives a poor approximation of a fixed point. Another disadvantage is that it is difficult to obtain the adjacent simplex to a given one when a certain vertex is dropped, and even harder to find the new vertex. Shapley has given an algorithm [59] based on this triangulation, but there is little to recommend its use.

3.11 Examples (n = 2).

$\frac{1}{4} K_1$

$k_1((\frac{1}{4},\frac{1}{2})^T,(1,2))$ $k_1((\frac{1}{2},0)^T,(2,1))$

$\frac{1}{4} J_1$

(Central vertices are heavy dots)

$j_1((\frac{1}{4},\frac{1}{4})^T,(1,2),(+1,-1)^T)$

$j_1((\frac{3}{4},\frac{1}{4})^T,(2,1),(-1,+1)^T)$

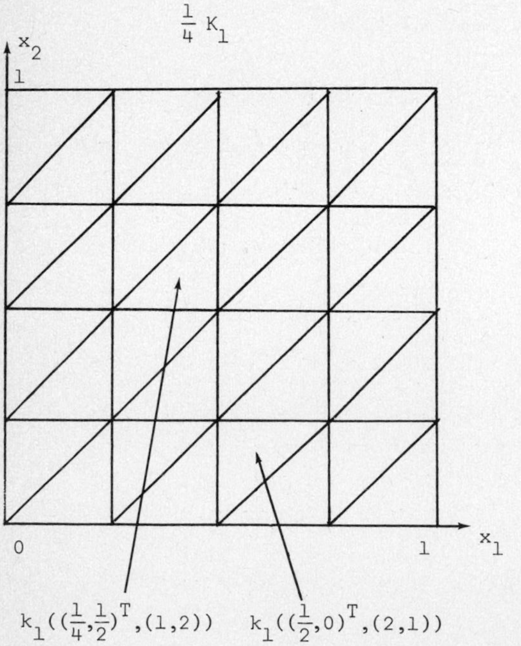

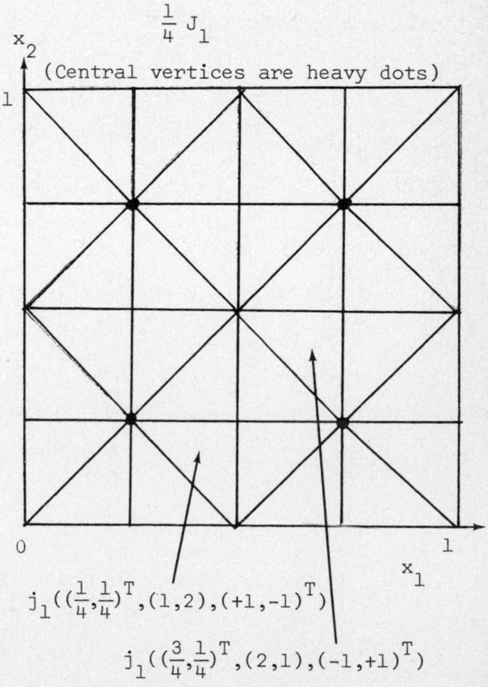

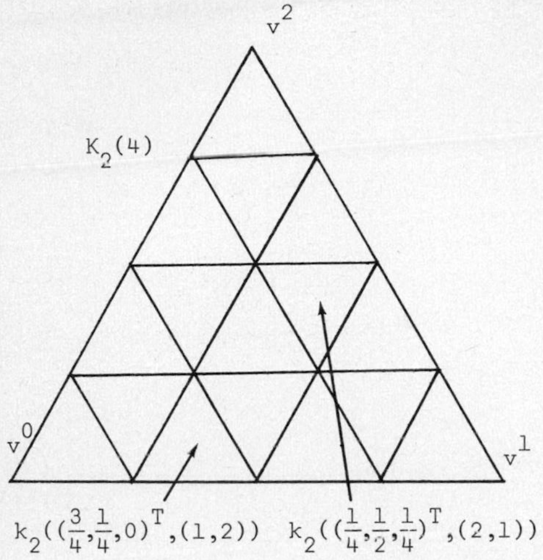

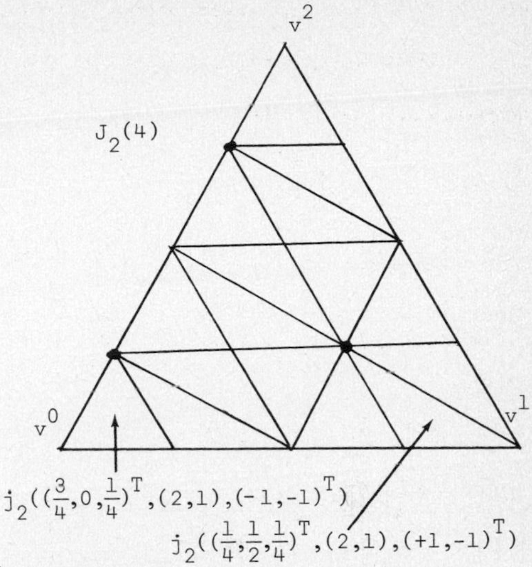

III.4 Pivot Rules. All our algorithms will generate a sequence of simplices $\sigma_1,\sigma_2,\ldots$ of a triangulation G with σ_i and σ_{i+1} adjacent; we obtain σ_{i+1} from σ_i by dropping a selected vertex y^- of σ_i and adding a new vertex y^+. The rules for obtaining σ_{i+1} from σ_i and y^- are called the pivot rules of G. Since typically hundreds of these "pivots" are necessary to obtain a sufficiently accurate fixed point, it is obvious that the pivot rules should be simple.

4.1 K_1. Let $\sigma = \langle y^0,\ldots,y^n\rangle = k_1(y^0,\pi)$ be given. We wish to obtain $\tau = \langle z^0,\ldots,z^n\rangle = k_1(z^0,\rho)$, with all vertices of σ except y^i vertices of τ. The table below shows how z^0 and ρ depend on y^0, π, and i -- from this table it is easy to obtain each z^i and in particular the new vertex of τ.

	z^0	ρ
$i = 0$	$y^0 + u^{\pi(1)}$	$(\pi(2),\ldots,\pi(n),\pi(1))$
$0 < i < n$	y^0	$(\pi(1),\ldots,\pi(i+1),\pi(i),\ldots,\pi(n))$
$i = n$	$y^0 - u^{\pi(n)}$	$(\pi(n),\pi(1),\ldots,\pi(n-1))$

For δK_1 replace $u^{\pi(1)}$ and $u^{\pi(n)}$ by $\delta u^{\pi(1)}$ and $\delta u^{\pi(n)}$.

4.2 $K_2(m)$. Let $\sigma = k_2(y^0,\pi) = \langle y^0,\ldots,y^n\rangle$ and $\tau = k_2(z^0,\rho)$ contain all vertices of σ except y^i. Then z^0 and ρ are obtained from y^0, π, and i according to the table above, except that $m^{-1}q^{\pi(1)}$ and $m^{-1}q^{\pi(n)}$ replace $u^{\pi(1)}$ and $u^{\pi(n)}$ respectively.

4.3 J_1. Let $\sigma = j_1(y^0,\pi,s) = \langle y^0,\ldots,y^n\rangle$ and $\tau = j_1(z^0,\rho,t)$ contain all vertices of σ except y^i. Then z^0, ρ and t are obtained from y^0, π, s, and i as shown in the table below.

	z^0	ρ	t
$i = 0$	$y + 2s_{\pi(1)}u^{\pi(1)}$	π	$s - 2s_{\pi(1)}u^{\pi(1)}$
$0 < i < n$	y	$(\pi(1),\ldots,\pi(i+1),\pi(i),\ldots,\pi(n))$	s
$i = n$	y	π	$s - 2s_{\pi(n)}u^{\pi(n)}$

For δJ_1 replace $u^{\pi(1)}$ in the top left-hand corner by $\delta u^{\pi(1)}$.

4.4 $J_2(m)$. Let $\sigma = j_2(y^0,\pi,s) = \langle y^0,\ldots,y^n\rangle$ and $\tau = j_2(z^0,\rho,t)$ contain all vertices of σ except y^i. Then z^0, ρ, and t are obtained from y^0, π, s, and i according to the table above, except that $m^{-1}q^{\pi(1)}$ replaces $u^{\pi(1)}$ in the top left-hand corner.

4.5 Remarks. One can check that the simplices above do share the appropriate vertices. The pivot rules can be motivated by examining the strings of inequalities (*) and (**) of 3.4 and 3.5. Moving between adjacent simplices is equivalent to crossing their common facet. Thus two terms in (*) or (**) change position, or an extreme term reappears on the other extreme. Clearly, the tables above reflect this behavior.

The pivot rules for iterated barycentric subdivision have been described by Shapley [59].

III.5 Scarf's Primitive Sets. All our algorithms will be based on triangulations. However, the first fixed-point algorithm of this type used an alternative notion of Scarf [57,58], that of primitive set. We will see in the next chapter that the essential property of a triangulation is that of 2.3(d)(ii); if σ is an n-simplex of G, τ a facet of σ, and $\tau \not\subseteq \partial C$, then there is precisely one other n-simplex σ' of G with τ as a facet. The notion of primitive set also distinguishes sets of n+1 points (like the vertices of a simplex) satisfying a similar condition.

Let $\tilde{S}^n$ denote the set $\{x \in R^{n+1} | v^T x = 1, \; x \leq 2v\}$; $\tilde{S}^n$ is a closed n-simplex with vertices y^i, $i \in N_0$, where $y^i = 2v - (2n+1)v^i$. Pick points $y^{n+1},\ldots,y^k$ in S^n such that for all $i \in N_0$ and $\ell > n$, $m > n$ with $\ell \neq m$, $y_i^\ell \neq y_i^m$.

5.1 Definition. A set of n+1 points $\{y^\ell | \ell \in L\}$ from $\{y^0,\ldots,y^k\}$ form a primitive set if there is no m, $n < m \leq k$, with $y_i^m > \min_{\ell \in L} y_i^\ell$ for each $i \in N_0$.

5.2 Example (n = 2). We have $y^0 = (-3,2,2)^T$, $y^1 = (2,-3,2)^T$ and $y^2 = (2,2,-3)^T$. Let $y^3 = (.9,0,.1)^T$, $y^4 = (.1,.9,0)^T$, $y^5 = (0,.1,.9)^T$ and $y^6 = (.05,.5,.45)^T$, $y^7 = (.45,.05,.5)^T$, $y^8 = (.5,.45,.05)^T$. The primitive sets are $\{y^1,y^2,y^3\}$ etc., $\{y^0,y^4,y^6\}$, $\{y^0,y^5,y^6\}$ etc., $\{y^3,y^7,y^8\}$ etc. and $\{y^6,y^7,y^8\}$. (Here "etc." means other triples obtained by the permutation $(y^0 y^1 y^2)(y^3 y^4 y^5)(y^6 y^7 y^8)$.)

One can see the close relation of these primitive sets to the triangulation $K_2(2)$ by forming 2-simplices from each primitive set (see Scarf [58, chapter 7]):

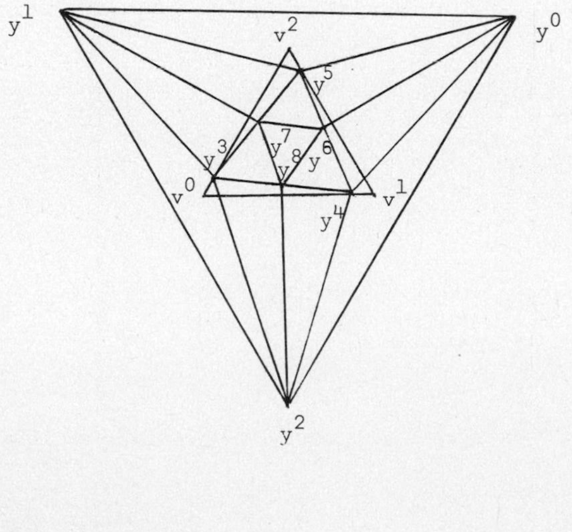

5.3 Lemma (Scarf [57]).

(a) Let $\{y^\ell | \ell \in L\}$ be a primitive set and $\ell^- \in L$. If $\{y^\ell | \ell \in L \sim \{\ell^-\}\} \not\subseteq \{y^i | i \in N_0\}$, then there is a unique ℓ^+ with $\{y^\ell | \ell \in (L \sim \{\ell^-\}) \cup \{\ell^+\}\}$ a primitive set.

(b) If $\{y^m | m \in M\} \subseteq \{y^i | i \in N_0\}$ and $|M| = n$, then there is a unique m^+ with $\{y^m | m \in M \cup \{m^+\}\}$ a primitive set. □

5.4 Remarks. In the next chapter we briefly describe how Scarf's algorithm uses primitive sets. Let us now compare primitive sets and triangulations. First, triangulations are essential for the more sophisticated algorithms we discuss in Chapter IX. However, primitive sets have the apparent advantage that the set of points $y^{n+1},\ldots,y^k$ determines the primitive sets, while there are many triangulations having these as vertices. Thus if a particular arrangement of vertices seems desirable, much less memory is required for primitive sets rather than triangulations. On the other hand, the replacement operation of 5.3(a) necessitates a search through $y^{n+1},\ldots,y^k$ unless these points are chosen with sufficient regularity. The almost universally used method of choosing these points leads to a direct correspondence between primitive sets and simplices of $K_2(m)$ for some $m > 0$. The method is due to Hansen; see Chapter 7 of Scarf [58].

III.6 Exercises.

6.1. Prove 1.3.

6.2. Let C be as in Section 2. Let G satisfy (i) and (iii) of 2.1, and (a) and (b) of 2.3. Show that G triangulates C.

6.3. Prove that J_1 triangulates R^n. Show that $x \in R^n$ lies in a simplex of J_1^+ of dimension less than n if and only if some x_i is an even integer, some $x_i - x_j$ is an even integer, or some $x_i + x_j$ is an even integer.

6.4. Prove that the simplices of $m^{-1}K_1$ (or of $m^{-1}J_1$) lying in $D^n = \{x \in R^n | 0 \leq x \leq u\}$ triangulates D^n, for any integer $m > 0$.

6.5. Prove that $J_2(m)$ triangulates S^n.

6.6. Prove 5.3.

6.7. Let $\langle x^0,y^1,\ldots,y^n\rangle$ and $\langle z^0,y^1,\ldots,y^n\rangle$ be non-intersecting n-simplices lying in an affine subspace of dimension n. Let $X = \begin{bmatrix} 1 & 1 & \cdots & 1 \\ x^0 & y^1 & \cdots & y^n \end{bmatrix}$ and $Z = \begin{bmatrix} 1 & 1 & \cdots & 1 \\ z^0 & y^1 & \cdots & y^n \end{bmatrix}$. Show that the determinants of X and Z have opposite signs.

CHAPTER IV: ALGORITHMS TO FIND COMPLETELY-LABELLED SIMPLICES

While the proof of Sperner's lemma given in I.4 does not suggest any method to find completely-labelled simplices, we noted in example I.4.3 that n-simplices of G with the labels $0,1,\ldots,n-1$ formed paths. Cohen [6] gave a proof of Sperner's lemma based on these paths; we present his argument in Section 1. Cohen's proof is inductive--we still have the problem of how to start. Two possible methods will be given in Section 2. Sections 3 and 4 show how these methods can be implemented using $K_2(m)$. Section 5 describes Scarf's algorithm, and Section 6 illustrates the algorithms with an example.

IV.1 The Graph Γ_n. Suppose we have a triangulation G of S^n with the vertices of G admissibly labelled, i.e., no vertex in S^n_i has the label i. We call members of G and G^{n-1} merely n-simplices and (n-1)-simplices. An n-simplex is completely-labelled (c.l.) if its vertices carry all labels in N_0, while an n- or (n-1)-simplex is almost completely labelled (a.c.l.) if its vertices carry all the labels $0,1,\ldots,n-1$.

We can now form the following graph.

1.1 Definition. Let the nodes of Γ_n be a.c.l. n-simplices and a.c.l. (n-1)-simplices lying in S^n_n. Two nodes of Γ_n are adjacent if one is a face of the other or if they share an a.c.l. face.

1.2 Example (n = 2).

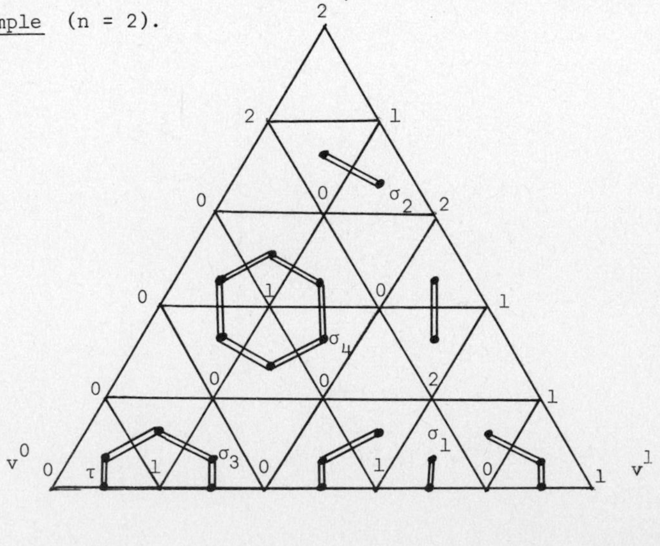

The nodes of Γ_2 are the heavy dots and its edges are the double lines. (The dot is placed on the barycenter of the corresponding simplex.)

We now analyze the degree of each node of Γ_n, i.e., the number of adjacent nodes.

Case 1. τ is an a.c.l. (n-1)-simplex in S_n^n. By III.2.3d(i), τ is a face of just one n-simplex σ and clearly σ is a.c.l. Hence τ has degree one. (See τ in 1.2.)

Case 2. σ is a c.l. n-simplex. Then σ has just one a.c.l. facet τ, that opposite the vertex of σ labelled n. If $\tau \subseteq S_n^n$, τ is a facet only of σ (III.2.3d(i)); σ is adjacent to τ and to no other node of Γ_n. If $\tau \not\subseteq S_n^n$, then since the labelling is admissible we have $\tau \not\subseteq \partial S^n$ and τ is a facet of exactly one other n-simplex σ' by III.2.3d(ii). Clearly σ' is a.c.l. Thus in either case σ has degree one. (See σ_1 and σ_2 in 1.2.)

Case 3. σ is an a.c.l. but not c.l. n-simplex. Since the n+1 vertices of σ have n labels, there is precisely one pair of vertices, y^1 and y^2, with the same label. Thus σ has precisely two a.c.l. faces τ_1 and τ_2 opposite y^1 and y^2, respectively. As in case 2, each τ_i either lies in S_n^n and is a node of Γ_n or leads to another a.c.l. n-simplex σ_i. In any case σ has degree two. (See σ_3 and σ_4 in 1.2.)

By combining these cases we immediately obtain

1.3 Theorem. Each connected component of Γ_n has one of the following forms:

(i) A simple circuit whose nodes are a.c.l. but not c.l. n-simplices.

(ii) A simple path whose intermediate nodes are a.c.l. but not c.l. n-simplices, and each of whose endpoints is either

(a) a c.l. n-simplex or

(b) an a.c.l. (n-1)-simplex in S_n^n. □

Example 1.2 shows that all four forms can arise.

Since the number of endpoints of a path is two, the total number of c.l. n-simplices and a.c.l. (n-1)-simplices in S_n^n is even. Clearly this proves the inductive step of Sperner's lemma.

IV.2 How can we find a c.l. n-simplex using Γ_n? First note that even if we have an a.c.l. (n-1)-simplex in S^n_n, tracing the path in Γ_n from this endpoint may take us back to S^n_n, as in the leftmost path in 1.2. The only way to be sure of avoiding this behavior is to have only one a.c.l. (n-1)-simplex in S^n_n.

The first method, to be discussed in Section 3, ensures that this is true by adding an artificial layer to S^n_n. The second method, described in Section 4, mimics the inductive proof by linking together all the graphs Γ_j, $j \in N$. Before we give details of these methods using $K_2(m)$, we can give an intuitive idea of their operation.

Suppose that we start with the labelled triangulation shown below. Γ_2 is also shown.

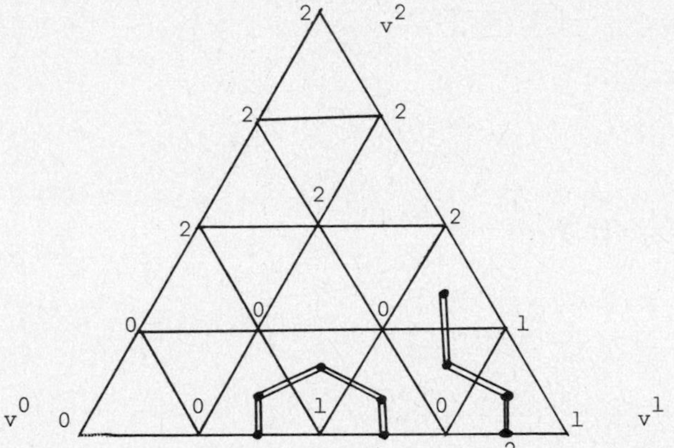

The first method adds an extra strip below S^2_2 and labels it to ensure a unique a.c.l. 1-simplex in the new boundary. The natural extension of Γ_2 is also shown. Note that there is now a path from the unique a.c.l. 1-simplex to a c.l. 2-simplex.

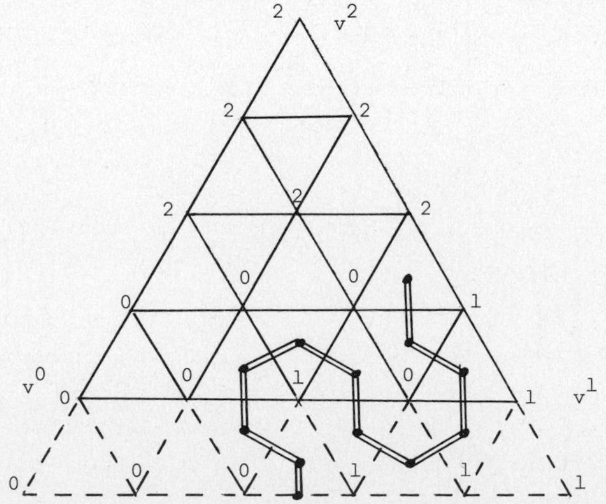

Now associate S^2_2 with S^1 and construct Γ_1:

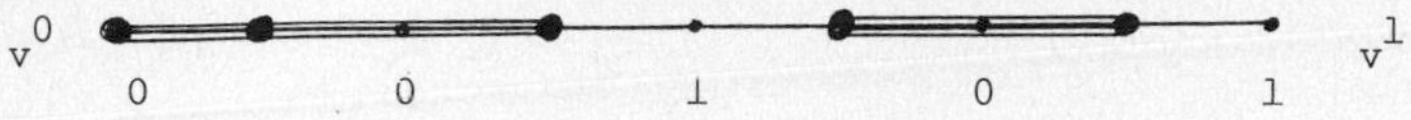

Then the union of Γ_1 and Γ_2 forms a graph Γ:

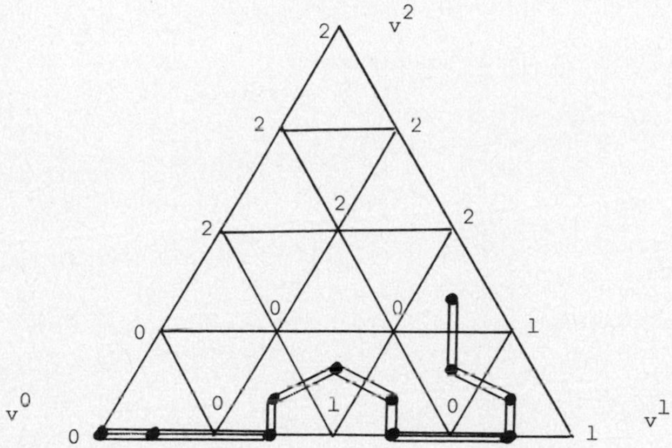

Γ contains a path from the only "a.c.l." 0-simplex in "S^1_1" -- v^0 --to a c.l. 2-simplex.

IV.3 The Artificial Start Algorithm (Kuhn [38]). We have a given function $f: S^n \to S^n$. We obtain an approximate fixed point of f by finding a completely labelled simplex of $K_2(m)$ using the standard labelling. To do this we first show how to extend $K_2(m)$ to a triangulation of S^n with an extra "layer" below S^n_n.

The arguments of III.3.6 show that the simplices of K_1 that meet $m\tilde{C}^n = \{x \in R^n | m \geq x_1 \geq \ldots \geq x_n \geq -1\}$ triangulate $m\tilde{C}^n$; hence those of $m^{-1}K_1$ that meet $\tilde{C}^n = \{x \in R^n | 1 \geq x_1 \geq \ldots \geq x_n \geq -m^{-1}\}$ triangulate $\tilde{C}^n$. By applying the homeomorphism h, we obtain a triangulation $\tilde{K}_2(m)$ of $\tilde{S}^n = \{x \in R^{n+1} | v^T x = 1,\ x_i \geq 0,$ $i = 0,\ldots,n-1,\ x_n \geq -m^{-1}\}$. $\tilde{S}^n$ is S^n with an extra layer added to S^n_n, as described informally in Section 2.

Each vertex of $\tilde{K}_2(m)$ lies in S^n or has its last coordinate $-m^{-1}$. If $y \in \tilde{K}^0_2(m)$ lies in S^n, we label it $\min\{j \in N_0 | f_j(y) \leq y_j > 0\}$; if $y \notin S^n$, we label it $\min\{j \in N_0 | y_j = \max_k y_k\}$. The latter labelling roughly corresponds to extending the function f to $\tilde{S}^n$ so that $f(x) = (n+1)^{-1}v$ for

$x \in \tilde{S}^n_n = \{x \in \tilde{S}^n | x_n = -m^{-1}\}$.

If we let $\tilde{S}^n_i = \{x \in \tilde{S}^n | x_i = 0\}$ for $i = 0,\dots,n-1$, we clearly have an admissible labelling, in the sense that no vertex in $\tilde{S}^n_i$ has the label i. Also, it is apparent that if $\sigma \in \tilde{K}_2(m)$ does not lie in S^n, then each of its vertices has n^{th} coordinate 0 or $-m^{-1}$ and cannot be labelled n. Hence all completely labelled simplices of $\tilde{K}_2(m)$ in fact lie in S^n and give approximate fixed points of f. We will therefore have the makings of an algorithm if we can establish that there is just one a.c.l. (n-1)-simplex τ^* in $\tilde{S}^n_n$.

Let $m = kn$ for some integer $k > 0$. Then we have

<u>3.1 Lemma</u>. There is just one a.c.l. (n-1)-simplex τ^* of $\tilde{K}^{n-1}_2(m)$ in $\tilde{S}^n_n$.

<u>Proof</u>. Note that the simplices of $\tilde{K}_2(m)$ are defined as in 3.7, except that each vertex must lie in $\tilde{S}^n$ but not necessarily in S^n. We first exhibit an a.c.l. (n-1)-simplex τ^* in $\tilde{S}^n_n$ and then show that it is unique.

Let $y^* = ((k+1)m^{-1},km^{-1},\dots,km^{-1},-m^{-1})^T$ and $\pi^* = (1,2,\dots,n-1)$. The (n-1)-simplex $\tau^* = \tilde{k}_2(y^*,\pi^*)$ is defined in the natural way. Its vertices are $y^0,\dots,y^{n-1}$ with $y^0 = y^*$ and $y^i = y^{i-1} + m^{-1}q^{\pi^*(i)}$ for $i = 1,2,\dots,n-1$. Thus $y^i = (km^{-1},\dots,km^{-1},(k+1)m^{-1},km^{-1},\dots,km^{-1},-m^{-1})^T$, with the $(k+1)m^{-1}$ in the ith coordinate. Then y^i has label i for each i. Also, τ^* is a facet of $\tilde{K}_2(y^*,(1,2,\dots,n))$. Hence τ^* is an a.c.l. (n-1)-simplex in $\tilde{S}^n_n$.

Now let $\tau = \langle\bar{y}^0,\dots,\bar{y}^{n-1}\rangle = \tilde{k}_2(\bar{y}^0,\bar{\pi})$ be an a.c.l. (n-1)-simplex in $\tilde{S}^n_n$. We must show $\tau = \tau^*$. Let $\bar{y}^i = (a_{i0}m^{-1},\dots,a_{i,n-1}m^{-1},-m^{-1})^T$ for each i. Then $\sum_{j=0}^{n-1} a_{ij} = m+1 = kn+1$ for each i. Hence $\bar{y}^i$ can have label j only if $a_{ij} \geq k+1$.

If $a_{00} < k+1$, each $a_{i0} < k+1$ and no vertex has label 0. If $a_{00} > k+1$, there is some j with $a_{0j} < k$, so that $a_{ij} \leq k$ for each i, and no vertex has label j. Thus $a_{00} = k+1$. A similar argument shows that no a_{0j} can exceed $k+1$, while if any $a_{0j} < k$ then $a_{i0} \geq k \geq a_{ij}$ for all i and no vertex has label j. In conclusion, $\bar{y}^0$ must be y^*.

If $\bar{\pi}(1) \neq 1$, then $\bar{y}^1$ has label 0--a contradiction. Similarly $\bar{\pi}(2)$ must be 2, etc., so that $\bar{\pi} = \pi^*$ and $\tau = \tau^*$. Thus τ^* is the only a.c.l. simplex of $\tilde{K}_2^{n-1}(m)$ lying in $\tilde{S}_n^n$. □

3.2 Algorithm. We are given $f: S^n \to S^n$. Pick $m = kn$ for some integer $k > 0$. Triangulate $\tilde{S}^n$ with $\tilde{K}_2(m)$, and label the vertices of $\tilde{K}_2(m)$ as above. Let τ^* be the (unique) a.c.l. (n-1)-simplex of $\tilde{K}_2^{n-1}(m)$ lying in $\tilde{S}_n^n$.

Step 0: Let σ_1 be the unique n-simplex of $\tilde{K}_2(m)$ that has τ^* as a facet. Let y^+ be the vertex of σ_1 that is not a vertex of τ^*. Set $\ell = 1$.

Step 1: Calculate the label of y^+. If it is n, STOP; σ_ℓ is completely labelled and yields an approximate fixed point of f. Otherwise, the label of y^+ duplicates that of exactly one other vertex of σ_ℓ, say y^-.

Step 2: Find the simplex $\sigma_{\ell+1}$ that has as vertices all the vertices of σ_ℓ except y^-. Let y^+ be the vertex of $\sigma_{\ell+1}$ that is not a vertex of σ_ℓ. Set $\ell \leftarrow \ell+1$ and return to Step 1.

3.3 Proof of Convergence. First we show that $\sigma_1, \sigma_2, \ldots$ are distinct. Assume the contrary, and let $\sigma_j = \sigma_\ell$, $j < \ell$ be the first duplication. If $j > 1$, then σ_j is adjacent to $\sigma_{j-1}, \sigma_{j+1}$ and $\sigma_{\ell-1}$ in the graph Γ_n. Since no node of Γ_n has degree greater than two, two of these must be equal. We cannot have $\sigma_{j-1} = \sigma_{j+1}$ or $\sigma_{j-1} = \sigma_{\ell-1}$, since these would be previous duplications. Thus $\sigma_{j+1} = \sigma_{\ell-1}$ and $j+1 = \ell-1$, i.e., $\ell = j+2$ and $\sigma_j = \sigma_{j+2}$. Now consider the vertex y of σ_{j+1} that is not a vertex of σ_j. Since the other vertex with the same label as y is dropped to obtain σ_{j+2}, we have that y is a vertex of σ_{j+2}. This is a contradiction. On the other hand, if $j = 1$, then σ_j is adjacent to σ_{j+1} and $\sigma_{\ell-1}$. But since σ_1 has degree one in Γ_n, we have $\sigma_{j+1} = \sigma_{\ell-1}$, yielding the same contradiction.

Since the simplices generated are distinct and there is a finite number of simplices of $\tilde{K}_2(m)$, the algorithm must terminate. If it terminates with σ_ℓ in Step 2 because $\sigma_{\ell+1}$ does not exist, then σ_ℓ has an a.c.l. facet in $\partial\tilde{S}^n$. Since the labelling is admissible, this facet must lie in $\tilde{S}_n^n$ and hence be τ^*. Thus $\sigma_\ell = \sigma_1$, contradicting the fact that all simplices are distinct.

Hence the algorithm terminates in Step 1 with a completely labelled simplex. □

This is the only time we give a proof of convergence for an algorithm; the proof is similar for any algorithm to be described. Usually we merely describe a graph with the same kind of property as Γ_n. The algorithm is then a formal statement of: "Start at a uniquely-defined simplex and trace a path in the graph until a desired simplex is found."

3.4 Remark. Lemke and Howson [45] first gave a proof of convergence based on the novel arguments above in their seminal paper on equilibrium points of bimatrix games. The uniqueness of the argument stems from its reliance on purely combinatorial reasoning to assure finite convergence--there is no monotonic evolution. Such proofs are now standard in complementary pivot theory. Lemke [44] has a 1970 survey on complementary pivot theory, while Gould and Tolle [26] and the author [63] discuss its general application from an abstract viewpoint.

IV.4 The Variable Dimension Algorithm. The version of this algorithm we will present is that of Kuhn [39]. Shapley constructed an algorithm independently [59], using barycentric subdivision. However, both algorithms are based on the pioneering work of Scarf [57] using primitive sets.

4.1 Definitions. Let $S^n_{(j+1)} = \{x \in S^n | x_{j+1} = \ldots = x_n = 0\}$ for $j = 0,1,\ldots,n-1$. $S^n_{(j+1)}$ is a closed j-simplex naturally corresponding to S^j. For $j \in N$, let Γ_j be the graph whose nodes are j-simplices in $S^n_{(j+1)}$ or (j-1)-simplices in $S^n_{(j)}$ whose vertices carry the labels $0,1,\ldots,j-1$. Two nodes are adjacent if one is a face of the other or if they have a common face whose vertices carry the labels $0,1,\ldots,j-1$.

For each $j \in N$, Γ_j satisfies a result similar to 1.3, where j replaces n, $S^n_{(j)}$ replaces S^n_n, and "a.c.l." ("c.l.") is interpreted to mean "whose vertices carry all the labels $0,1,\ldots,j-1$ $(0,1,\ldots,j)$."

Let Γ be the union of Γ_j, for all $j \in N$, i.e., the set of nodes (edges) of Γ is the union of the sets of nodes (edges) of all Γ_j. Then we have

4.2 Theorem. Each connected component of Γ is either a simple circuit or a simple path from v^0 to a c.l. n-simplex.

The proof follows from 1.3. $\square$

A natural triangulation to use is $K_2(m)$. We need to be able to describe simplices of $K_2^j(m)$ in $S^n_{(j+1)}$. Let π be a permutation of $\{1,2,\ldots,j\}$. Then, for $y^0 \in K_2^0(m)$, $k_2(y^0,\pi)$ denotes the j-simplex $\langle y^0,\ldots,y^j\rangle$ with $y^i = y^{i-1} + m^{-1}q^{\pi(i)}$ for $1 \leq i \leq j$, if all y^i lie in $S^n_{(j+1)}$. The unique (j+1)-simplex of $K_2^{j+1}(m)$ lying in $S^n_{(j+2)}$ with $k_2(y^0,\pi)$ as a facet is $k_2(y^0,\pi')$, where $\pi' = (\pi(1),\ldots,\pi(j),j+1)$.

We then have the following.

4.3 Algorithm. We are given a function $f: S^n \to S^n$. Triangulate S^n with $K_2(m)$ and label vertex y with $\min\{j \mid f_j(y) \leq y_j > 0\}$.

Step 0: Let $j = 1$, $y^0 = v^0$, $\sigma_1 = k_2(y^0,\pi)$ with $\pi = (1)$ the permutation of $\{1\}$. Let $y^+ = y^1 = v^0 + m^{-1}q^1$. Set $\ell = 1$.

Step 1: Calculate the label of y^+. If it is j, go to Step 3. Otherwise, it duplicates the label of some vertex y^- of σ_ℓ.

Step 2: Let τ be the facet of σ_ℓ opposite y^-. If $\tau \subseteq S^n_{(j)}$ go to Step 4. Otherwise, let $\sigma_{\ell+1}$ be the unique j-simplex in $S^n_{(j+1)}$ sharing the facet τ with σ_ℓ. Let y^+ be the new vertex of $\sigma_{\ell+1}$. Set $\ell \leftarrow \ell+1$ and return to Step 1.

Step 3: (Increasing the dimension) We have the j-simplex $\sigma_\ell = k_2(y^0,\pi)$, say, with π a permutation of $\{1,2,\ldots,j\}$. The vertices of σ_ℓ have all the labels $0,1,\ldots,j$. If $j = n$, STOP; σ_ℓ is completely labelled and yields an approximate fixed point of f. Otherwise, let $\sigma_{\ell+1} = k_2(y^0,\pi')$, where $\pi' = (\pi(1),\ldots,\pi(j),j+1)$, and let y^+ be the new vertex of $\sigma_{\ell+1}$. Set $\ell \leftarrow \ell+1$, $j \leftarrow j+1$ and return to Step 1.

Step 4: (Decreasing the dimension) We have the j-simplex $\sigma_\ell = k_2(y^0,\pi) = \langle y^0,\ldots,y^j\rangle$ with π a permutation of $\{1,2,\ldots,j\}$. σ_ℓ has a facet τ in $S^n_{(j)}$, and it is clear that τ must be $\langle y^0,\ldots,y^{j-1}\rangle$ and $\pi(j) = j$. The vertices of τ have all the labels $0,1,\ldots,j-1$. Let $\sigma_{\ell+1} = k_2(y^0,\pi')$ with $\pi' = (\pi(1),\ldots,\pi(j-1))$, and let y^- be the vertex of $\sigma_{\ell+1}$ with label $j-1$. Set $\ell \leftarrow \ell+1$, $j \leftarrow j-1$ and return to Step 2.

This algorithm produces a sequence of simplices of varying dimensions following the path of Γ from v^0 to a c.l. n-simplex.

Alternatively, we can "fill out" each simplex of dimension less than n by adding artificial vertices. We then obtain a triangulation of $\tilde{S}^n =$ $\{x \in R^{n+1} | v^T x = 1,\ x \leq 2v\}$. (See III.5.) The new vertex $y^i = 2v - (2n+1)v^i$ is labelled $i-1$ (mod $n+1$). The algorithm is then equivalent to that of Section 3. Using the example of Section 2, we obtain the extended triangulation of S^n shown below. The natural extension of Γ_2 is also shown, and its close relationship with $\Gamma = \Gamma_1 \cup \Gamma_2$ is clear.

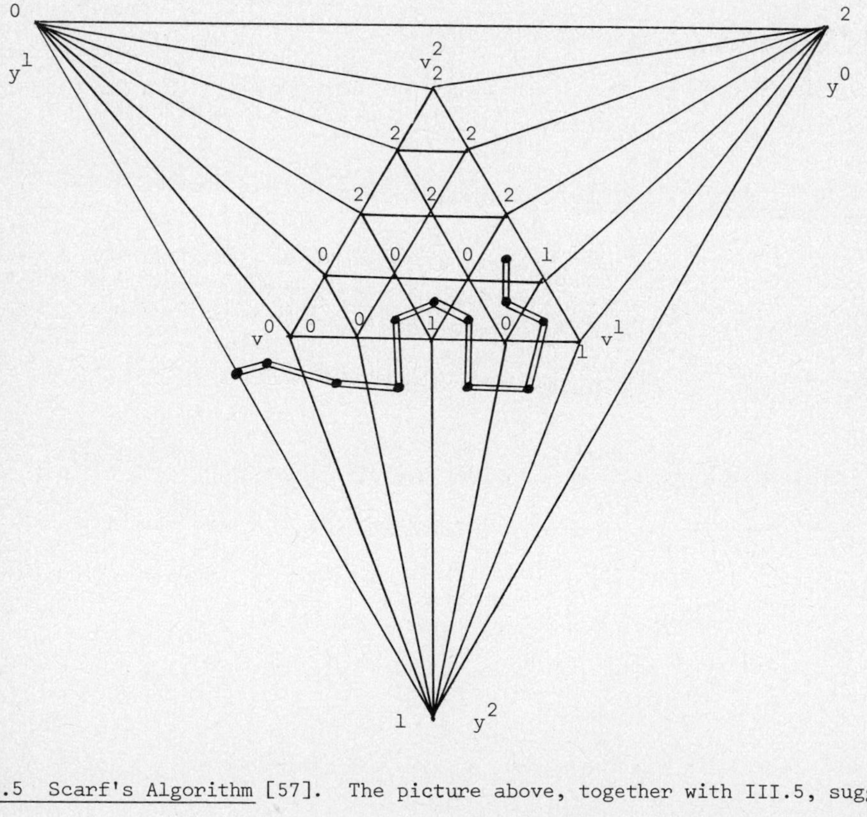

<u>IV.5 Scarf's Algorithm</u> [57]. The picture above, together with III.5, suggests how an algorithm using primitive sets can be constructed. Let $\tilde{S}^n$, y^i, $i \in N_0$ be as in the last section. Label y^i $i-1$ mod $n+1$, for $i \in N_0$, and label y^ℓ $\min\{j | f_j(y^\ell) \leq y_j^\ell > 0\}$ if $\ell > n$. Start with the unique primitive set including $\{y^i | i \in N\}$ given by III.5.3(b). Proceed as in 3.2, using the replacement step of III.5.3(a) in each Step 2.

Scarf's algorithm used a different labelling scheme: label y^i i for each $i \in N_0$ and label y^ℓ $\min\{j \mid f_j(y^\ell) \geq y_j^\ell\}$ for $\ell > n$. Again a set of close points carrying all labels gives an approximate fixed point, and the algorithm is the same. Vertgeim [71] discusses some general labelling rules, including both those used here.

IV.6 Examples. We take $n = 2$ so that the progress of iterations can be seen on a diagram. We execute the algorithms using the formal descriptions to show how rapidly and conveniently the operations can be performed using K_2.

Define $f: S^2 \to S^2$ by $f(x) = \begin{bmatrix} .2 & .1 & .3 \\ .3 & .4 & .3 \\ .5 & .5 & .4 \end{bmatrix} x$. We will use $K_2(4)$ to find an approximate fixed point of f. In fact, f has just one fixed point, $x^* = (7,11,15)^T/33$.

6.1 The Artificial Start Algorithm Illustrated. We have $m = 4 = kn = 2.2$. The course of the algorithm is shown in the table below. The operations are performed from left to right and top to bottom. We omit the denominator of 4 in every vertex.

The picture below shows the simplices generated.

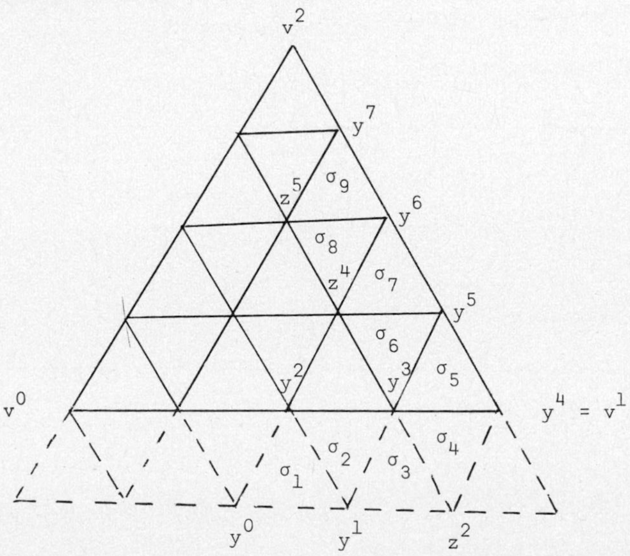

ℓ	σ_ℓ	y^+	$f(y^+)$	Label of y^+	y^-	Table III.4.1 Case
1	$\langle y^0,y^1,y^2\rangle = \left\langle \begin{pmatrix}3\\2\\-1\end{pmatrix}, \begin{pmatrix}2\\3\\-1\end{pmatrix}, \begin{pmatrix}2\\2\\0\end{pmatrix} \right\rangle$ Labels 0 1 ? $\tilde{k}_2(y^0,\pi^0)$, $\pi^0 = (1,2)$	$\begin{pmatrix}2\\2\\0\end{pmatrix}$	$\begin{pmatrix}.15\\.35\\.5\end{pmatrix}$	0	y^0	$i = 0$
2	$\langle y^1,y^2,y^3\rangle = \left\langle \begin{pmatrix}2\\3\\-1\end{pmatrix}, \begin{pmatrix}2\\2\\0\end{pmatrix}, \begin{pmatrix}1\\3\\0\end{pmatrix} \right\rangle$ Labels 1 0 ? $\tilde{k}_2(y^1,\pi^1)$, $\pi^1 = (2,1)$	$\begin{pmatrix}1\\3\\0\end{pmatrix}$	$\begin{pmatrix}.125\\.375\\.5\end{pmatrix}$	0	y^2	$0 < i < n$
3	$\langle y^1,z^2,y^3\rangle = \left\langle \begin{pmatrix}2\\3\\-1\end{pmatrix}, \begin{pmatrix}1\\4\\-1\end{pmatrix}, \begin{pmatrix}1\\3\\0\end{pmatrix} \right\rangle$ Labels 1 ? 0 $\tilde{k}_2(y^1,\pi^2)$, $\pi^2 = (1,2)$	$\begin{pmatrix}1\\4\\-1\end{pmatrix}$	?	1	y^1	$i = 0$
4	$\langle z^2,y^3,y^4\rangle = \left\langle \begin{pmatrix}1\\4\\-1\end{pmatrix}, \begin{pmatrix}1\\3\\0\end{pmatrix}, \begin{pmatrix}0\\4\\0\end{pmatrix} \right\rangle$ Labels 1 0 ? $\tilde{k}_2(z^2,\pi^3)$, $\pi^3 = (2,1)$	$\begin{pmatrix}0\\4\\0\end{pmatrix}$	$\begin{pmatrix}.1\\.4\\.5\end{pmatrix}$	1	z^2	$i = 0$
5	$\langle y^3,y^4,y^5\rangle = \left\langle \begin{pmatrix}1\\3\\0\end{pmatrix}, \begin{pmatrix}0\\4\\0\end{pmatrix}, \begin{pmatrix}0\\3\\1\end{pmatrix} \right\rangle$ Labels 0 1 ? $\tilde{k}_2(y^3,\pi^4)$, $\pi^4 = (1,2)$	$\begin{pmatrix}0\\3\\1\end{pmatrix}$	$\begin{pmatrix}.15\\.375\\.475\end{pmatrix}$	1	y^4	$0 < i < n$
6	$\langle y^3,z^4,y^5\rangle = \left\langle \begin{pmatrix}1\\3\\0\end{pmatrix}, \begin{pmatrix}1\\2\\1\end{pmatrix}, \begin{pmatrix}0\\3\\1\end{pmatrix} \right\rangle$ Labels 0 ? 1 $\tilde{k}_2(y^3,\pi^5)$, $\pi^5 = (2,1)$	$\begin{pmatrix}1\\2\\1\end{pmatrix}$	$\begin{pmatrix}.175\\.35\\.475\end{pmatrix}$	0	y^3	$i = 0$

ℓ	σ_ℓ	y^+	$f(y^+)$	Label of y^+	y^-	Table III.4.1 Case
7	$\langle z^4,y^5,y^6\rangle = \left\langle \begin{pmatrix}1\\2\\1\end{pmatrix}, \begin{pmatrix}0\\3\\1\end{pmatrix}, \begin{pmatrix}0\\2\\2\end{pmatrix} \right\rangle$ Labels 0 1 ? $\tilde{k}_2(z^4,\pi^6)$, $\pi^6 = (1,2)$	$\begin{pmatrix}0\\2\\2\end{pmatrix}$	$\begin{pmatrix}.2\\.35\\.45\end{pmatrix}$	1	y^5	$0 < i < n$
8	$\langle z^4,z^5,y^6\rangle = \left\langle \begin{pmatrix}1\\2\\1\end{pmatrix}, \begin{pmatrix}1\\1\\2\end{pmatrix}, \begin{pmatrix}0\\2\\2\end{pmatrix} \right\rangle$ Labels 0 ? 1 $\tilde{k}_2(z^4,\pi^7)$, $\pi^7 = (2,1)$	$\begin{pmatrix}1\\1\\2\end{pmatrix}$	$\begin{pmatrix}.225\\.325\\.45\end{pmatrix}$	0	z^4	$i = 0$
9	$\langle z^5,y^6,y^7\rangle = \left\langle \begin{pmatrix}1\\1\\2\end{pmatrix}, \begin{pmatrix}0\\2\\2\end{pmatrix}, \begin{pmatrix}0\\1\\3\end{pmatrix} \right\rangle$ Labels 0 1 ? $\tilde{k}_2(z^5,\pi^8)$, $\pi^8 = (1,2)$	$\begin{pmatrix}0\\1\\3\end{pmatrix}$	$\begin{pmatrix}.25\\.325\\.425\end{pmatrix}$	2		

σ_9 is a completely labelled simplex. Note that in fact x^* lies in σ_8.

6.2 The Variable-Dimension Algorithm Illustrated. We use $K_2(4)$. The table shows the course of iterations, as does the diagram following.

ℓ	σ_ℓ	y^+	$f(y^+)$	Label of y^+	y^-	Table III.4.1 Case
1	$\langle y^0,y^1\rangle = \left\langle \begin{pmatrix}4\\0\\0\end{pmatrix},\begin{pmatrix}3\\1\\0\end{pmatrix}\right\rangle$ Labels 0 ? $k_2(y^0,\pi^0)$, $\pi^0 = (1)$	$\begin{pmatrix}3\\1\\0\end{pmatrix}$	$\begin{pmatrix}.175\\.325\\.5\end{pmatrix}$	0	y^0	i = 0
2	$\langle y^1,y^2\rangle = \left\langle \begin{pmatrix}3\\1\\0\end{pmatrix},\begin{pmatrix}2\\2\\0\end{pmatrix}\right\rangle$ Labels 0 ? $k_2(y^1,\pi^1)$, $\pi^1 = (1)$	$\begin{pmatrix}2\\2\\0\end{pmatrix}$	$\begin{pmatrix}.15\\.35\\.5\end{pmatrix}$	0	y^1	i = 0
3	$\langle y^2,y^3\rangle = \left\langle \begin{pmatrix}2\\2\\0\end{pmatrix},\begin{pmatrix}1\\3\\0\end{pmatrix}\right\rangle$ Labels 0 ? $k_2(y^2,\pi^2)$, $\pi^2 = (1)$	$\begin{pmatrix}1\\3\\0\end{pmatrix}$	$\begin{pmatrix}.125\\.375\\.5\end{pmatrix}$	0	y^2	i = 0
4	$\langle y^3,y^4\rangle = \left\langle \begin{pmatrix}1\\3\\0\end{pmatrix},\begin{pmatrix}0\\4\\0\end{pmatrix}\right\rangle$ Labels 0 ? $k_2(y^3,\pi^3)$, $\pi^3 = (1)$	$\begin{pmatrix}0\\4\\0\end{pmatrix}$	$\begin{pmatrix}.1\\.4\\.5\end{pmatrix}$	1	-	- Step 2 of 4.3
5	$\langle y^3,y^4,y^5\rangle = \left\langle \begin{pmatrix}1\\3\\0\end{pmatrix},\begin{pmatrix}0\\4\\0\end{pmatrix},\begin{pmatrix}0\\3\\1\end{pmatrix}\right\rangle$ Labels 0 1 ? $k_2(y^3,\pi^4)$, $\pi^4 = (1,2)$	$\begin{pmatrix}0\\3\\1\end{pmatrix}$	$\begin{pmatrix}.15\\.375\\.475\end{pmatrix}$	1	y^4	$0 < i < n$

Same as in 6.1 from here on.

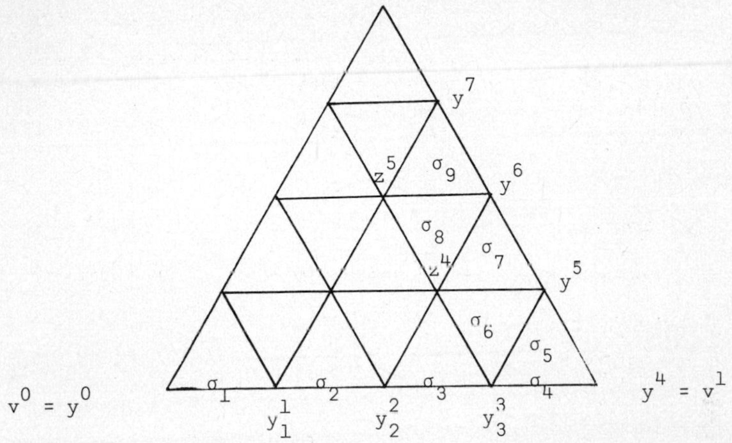

IV.7 Exercises

7.1. Apply the algorithms of Sections 3 and 4 to the labelled triangulation in Section 2.

7.2. Devise an artificial-start algorithm using a triangulation based on $J_2(m)$, where $m = kn$ and k is a positive even integer. Apply your algorithm to the example of Section 6.

7.3. Devise a variable-dimension algorithm using $J_2(m)$. Apply it to the example of Section 6.

7.4. We have not shown exactly how the triangulation at the end of Section 4 was obtained. Let $\tilde{S}^n = \{x \in R^{n+1} | v^T x = 1, \ x \leq 2v\}$ and $y^i = 2v - (2n+1)v^i$ for $i \in N_0$. If I,J is a partition of N_0 with $I \neq \emptyset$, let σ_I be the n-simplex with vertices v^i, $i \in I$, and y^j, $j \in J$. Let H be the set of all such σ_I. Prove that H is a triangulation. (Hint: do not use the methods of Chapter III. Consider the natural triangulation of the octahedron in R^{n+1}: $\{x \in R^{n+1} | \sum_{N_0} |x_i| = 1\}$.) Next show how any triangulation G of S^n can be extended to a triangulation $\tilde{G}$ of $\tilde{S}^n$. Prove that $\tilde{G}$ is a triangulation.

7.5. Refer to I.5.8, III.6.4. Triangulate D^n with $m^{-1}K_1$ or $m^{-1}J_1$. Label a vertex y i if $y \in C_i'$. Construct an algorithm to give a completely-labelled simplex. (Hint: find a set of n labels so that, independent of f, there is a unique (n-1)-simplex in the boundary of D^n with these labels.)

CHAPTER V: EXTENSIONS OF BROUWER'S THEOREM

Chapter II makes clear the need for more general fixed-point theorems and algorithms to approximate the resulting fixed points. Two extensions are necessary: one to relax the continuity of the function involved, and one to relax the requirements on the domain and range of the function. Section 1 discusses upper semi-continuity of point-to-set mappings. Section 2 introduces the fundamental concept of piecewise-linear approximation and with this tool proves Kakutani's theorem on the existence of fixed points for upper semi-continuous mappings. Section 3 extends Kakutani's theorem in ways that will be useful in applications.

V.1 Upper Semi-Continuity. Let us motivate our extension by considering a simple one-dimensional optimization problem. If we want to minimize $\theta_1(x) = x^2 - x$, we can form $f_1(x) = x - \nabla\theta_1(x) = x - (2x - 1) = 1 - x$. Then $f_1: [0,1] \to [0,1]$ is continuous and a fixed point gives the minimizer $x^* = \frac{1}{2}$ of θ_1. Now let $\theta_2(x) = \frac{1}{2}|x - \frac{1}{2}|$. $\nabla\theta_2(x)$ is $-\frac{1}{2}$ for $x < \frac{1}{2}$, $\frac{1}{2}$ for $x > \frac{1}{2}$ and undefined for $x = \frac{1}{2}$. Thus $f_2(x) = x - \nabla\theta_2(x)$ is $\frac{1}{2} + x$ for $x < \frac{1}{2}$ and $x - \frac{1}{2}$ for $x > \frac{1}{2}$. f_2 has no fixed points. Clearly, we must define $f_2(\frac{1}{2})$ somehow, and thus extend the notion of differentiability.

1.1 Definition. Let $\theta: R^n \to R$ be convex. Then $h \in R^n$ is a subgradient of θ at x if $h^T(z-x) \leq \theta(z) - \theta(x)$ for all $z \in R^n$. The subdifferential of θ at x, denoted $D\theta(x)$, is the set of all subgradients of θ at x.

It follows from separation theorems that $D\theta(x)$ is nonempty for all x when θ is convex, and clearly $D\theta(x)$ is convex. If θ is differentiable at x, $D\theta(x) = \{\nabla\theta(x)\}$, and $0 \in D\theta(x)$ if and only if x minimizes θ. The subdifferential is therefore a natural extension of the gradient.

Since $D\theta$ is set-valued, we are led to consider point-to-set mappings. For our example, we have $D\theta_2(x) = \{-\frac{1}{2}\}$ if $x < \frac{1}{2}$, $D\theta_2(\frac{1}{2}) = [-\frac{1}{2}, +\frac{1}{2}]$ and $D\theta_2(x) = \{+\frac{1}{2}\}$ if $x > \frac{1}{2}$. If we define $F(x) = \{x\} - D\theta_2(x)$, we have F shown by the diagram below:

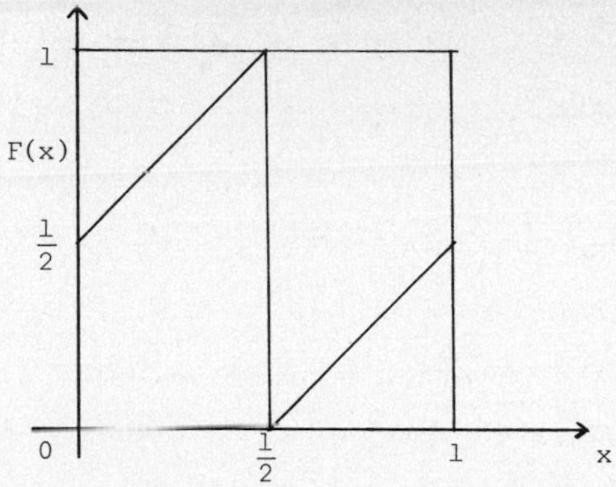

We naturally call $\frac{1}{2}$ a fixed point of F, meaning $\frac{1}{2} \in F(\frac{1}{2})$.

Clearly, we need some conditions on F to guarantee fixed points. Some sort of continuity is required, and motivated by the example above we see that we can allow F(x) to suddenly increase but not suddenly decrease.

1.2 Definition. Denote by $P(R^m)$ the set of all subsets of R^m. Let $C \subseteq R^m$ and $F: C \to P(R^p)$ be a point-to-set mapping. Then we say F is upper semi-continuous (u.s.c.) if

(i) for all $x \in C$, $F(x)$ is compact; and

(ii) for all $x \in C$, for all $\varepsilon > 0$, there is a $\delta > 0$ such that if $z \in B(x,\delta) \cap C$, $F(z) \subseteq B(F(x),\varepsilon)$.

If f is a function from C to R^p, then $F: C \to P(R^p)$ with $F(x) = \{f(x)\}$ is a point-to-set mapping--we write $F = \{f\}$. Note that f is continuous iff $\{f\}$ is u.s.c.

The most important property of u.s.c. mappings (often taken as a weaker definition of upper semi-continuity) is the following:

1.3 Lemma. If $F: C \to P(R^p)$ is u.s.c., $\{x^k\}$ is a sequence of points in C tending to x^*, and $\{y^k\}$ is a sequence of points tending to y^* with $y^k \in F(x^k)$ for each k, then $y^* \in F(x^*)$.

Proof. For any $\varepsilon > 0$, we can find k such that $\|y^* - y^k\|_2 \leq \varepsilon/2$ and $F(x^k) \subseteq B(F(x^*), \varepsilon/2)$. Thus $y^* \in B(F(x^*), \varepsilon)$ for every $\varepsilon > 0$. Since $F(x^*)$ is compact, $y^* \in F(x^*)$. □

Clearly, that F is u.s.c. from C to $P(C)$ for some compact convex $C \subseteq R^n$ does not suffice for F to have a fixed point. Indeed, the map that takes every point to the empty set is u.s.c. Also, if $F(\frac{1}{2})$ in the example above 1.2 is changed from $[0,1]$ to $\{0,1\}$, then F remains u.s.c. but has no fixed point. These examples motivate the restriction that the values of F must be nonempty convex sets. If F is not convex-valued, then it can be "convexified". This is one part of the following theorem, which collects all the results we need about composing and combining u.s.c. maps.

1.4 Theorem. In all parts below, unless otherwise specified, all u.s.c. mappings are from $C \subseteq R^m$ to $P(R^p)$.

(a) If F is u.s.c. and $D \subseteq C$ is compact, so is $F(D)$.

(b) Let F be u.s.c. and $G: D \subseteq R^p \to P(R^\ell)$ be u.s.c. with $F(C) \subseteq D$. Then $H = GF: C \to P(R^\ell)$ is u.s.c., where $H(x) = G(F(x))$.

(c) Let F_i be u.s.c., $i = 1,2,\ldots,k$ and $F: C \to P(R^p)$ be defined by $F(x) = \bigcup_{i=1}^k F_i(x)$. Then F is u.s.c.

(d) Let F_i be u.s.c., $i = 1,2,\ldots,k$, and $F: C \to P(R^p)$ be defined by $F(x) = F_1(x) + \ldots + F_k(x)$. Then F is u.s.c.

(e) Let F be u.s.c. and C be closed with $C \subseteq D$. Then $G: D \to P(R^p)$ is u.s.c., where $G(x) = F(x)$ if $x \in C$, ϕ if $x \in D \sim C$.

(f) If F is u.s.c., then $\text{conv}\, F: C \to P(R^p)$ is u.s.c., where $(\text{conv}\, F)(x) = \text{conv}(F(x))$.

(g) If $F_i: C \subseteq R^n \to P(R^{p_i})$ is u.s.c., $i = 1,2,\ldots,k$, and $p = \sum p_i$, then F is u.s.c., where $F(x) = F_1(x) \times \ldots \times F_k(x)$.

Proof. For (a), (b), (c) and (g), see Theorems 3, 1', 3' and 4' (pages 116-120) of Berge [3]. Part (d) follows from (g) and (b), with $G(z^1,z^2,\ldots,z^k) = z^1 + \ldots + z^k$. Part (e) is trivial from 1.2. We now prove (f).

Given $x \in C$ and $\varepsilon > 0$, determine $\delta > 0$ so that $z \in B(x,\delta) \cap C$ implies $F(z) \subseteq B(F(x),\varepsilon))$. Let $g \in \text{conv}\, F(z)$, so that $g = \sum_{i=1}^k \lambda_i g^i$, $\lambda_i \geq 0$ and $\sum_{i=1}^k \lambda_i = 1$ for some $g^i \in F(z)$, $i = 1,\ldots,k$. For each g^i, there is an $f^i \in F(x)$ with $\|g^i - f^i\|_2 \leq \varepsilon$. Since the Euclidean norm is a convex function,

$\|g - f\|_2 \leq \varepsilon$, where $f = \sum_{i=1}^{k} \lambda_i f^i \in \text{conv } F(x)$. Hence $\text{conv } F(z) \subseteq B(\text{conv } F(x),\varepsilon)$. Clearly, conv $F(x)$ is compact for any $x \in C$, and thus conv F is u.s.c. □

We generally use 1.4 as follows. In different (closed) regions, we have different u.s.c. maps (often functions) corresponding to a modification of the given point. These maps are extended to R^m by 1.4(e) and joined together by 1.4(c), and the result "convexified" by (f).

It is convenient to denote by R^{m*} the set of all nonempty convex subsets of R^m.

V.2 Kakutani's Theorem and Piecewise Linear Approximations. In 1941 Kakutani [32] generalized Brouwer's theorem to u.s.c. maps. He proved

2.1 Theorem. Let C be a compact convex subset of R^m, and $F: C \to C^*$ be u.s.c. Then F has a fixed point x^*, i.e., $x^* \in F(x^*)$.

To prove 2.1 we use the fundamental notion of a piecewise linear approximation to F.

2.2 Definition. Let $C \subseteq R^m$ be convex, with $\dim C = n$, and let G be a triangulation of C. Let $F: C \to P(R^p)$ be a point-to-set mapping whose values are nonempty. For each vertex y of G, pick $f(y) \in F(y)$. Each $x \in C$ lies in a unique simplex of G^+. Thus we can write $x = \sum_{i\in N_0} \lambda_i y^i$, $\lambda_i \geq 0$, $\sum_{i\in N_0} \lambda_i = 1$, where $\langle y^0,\ldots,y^n\rangle \in G$ and the pairs λ_i, y^i for $\lambda_i > 0$ are unique. Then set $f(x) = \sum_{i\in N_0} \lambda_i f(y^i)$. The function $f: C \to R^p$ is thus well-defined; we call f a piecewise linear (p.l.) approximation to F with respect to G.

2.3 Lemma. A piecewise linear approximation is a continuous function.

Proof. An easy exercise using the local finiteness of G. □

Proof of 2.1. We will follow Eaves' proof [12], which is closely based on Kakutani's original proof. If $\dim C = n < m$, we can map aff(C) homeomorphically onto R^n. We can therefore assume without loss of generality that $m = n$. Pick $c \in \text{int } C$. Since C is compact, we can embed it in a closed n-simplex S. Extend

F to S as follows. Let $F_1(x) = F(x)$ if $x \in C$, $F_1(x) = \emptyset$ if $x \in S \sim C$. Let $F_2(x) = \{c\}$ if $x \in S \sim \text{int } C$, $\emptyset$ if $x \in \text{int } C$. Finally, let $F_0(x) = \text{conv}(F_1(x) \cup F_2(x))$. By 1.4(e), (c) and (f), $F_0 \colon S \to S^*$ is u.s.c.

Further, if x^* is a fixed point of F_0, then certainly $x^* \in C$ (otherwise, $x^* \in F_0(x^*) = \{c\} \subseteq C$, a contradiction). If $x^* \in \text{int } C$, then $x^* \in F_0(x^*) = F_1(x^*) = F(x^*)$. So assume $x^* \in \partial C$. Then there is $f \in F_1(x^*) = F(x^*)$ and $\lambda \in [0,1]$ with $x^* = \lambda c + (1-\lambda)f$, since $F(x^*)$ is convex. If $\lambda > 0$, then since $c \in \text{int } C$ and $f \in C$, $x^* \in \text{int } C$; thus $\lambda = 0$, $x^* = f$, and $x^* \in F(x^*)$. Hence x^* is a fixed point of F.

We now prove that F_0 has a fixed point. Let G_k, $k = 1,2,\ldots$, be a sequence of triangulations of S with mesh $G_k \to 0$. (Such triangulations exist; for example, $K_2(m)$ can be mapped by a function taking S^n onto S barycentrically.) For each k, let f_k be a p.l. approximation to F_0 with respect to G_k. By 2.3 and Brouwer's theorem, each f_k has a fixed point x^k. Hence

$$x^k = \textstyle\sum_{i\in N_0} \lambda_{k,i} y^{k,i} = \sum_{i\in N_0} \lambda_{k,i} f^{k,i} \quad \text{for } \lambda_{k,i} \geq 0, \ \sum_{i\in N_0} \lambda_{k,i} = 1 \qquad (*)$$

with $\langle y^{k,0},\ldots,y^{k,n}\rangle \in G_k$, $f^{k,i} \in F_0(y^{k,i})$ for each $i \in N_0$.

As $k \to \infty$, x^k, all the $\lambda_{k,i}$'s and $f^{k,i}$'s remain in compact sets C, $[0,1]$ and C. Thus there is a subsequence (without loss of generality, the whole sequence) along which $x^k \to x^*$, $\lambda_{k,i} \to \lambda_i$ and $f^{k,i} \to f^i$. Since mesh $G_k \to 0$, we have that $y^{k,i} \to x^*$ for each $i \in N_0$. Because F_0 is u.s.c., we deduce from 1.3 that $f^i \in F_0(x^*)$ for each $i \in N$. Taking limits in (*), we have $x^* = \sum_{i\in N_0} \lambda_i f^i$, $\lambda_i \geq 0$, $\sum_{i\in N_0} \lambda_i = 1$.

But $F_0(x^*)$ is convex; hence x^* lies in $F_0(x^*)$ and the result is proved. □

2.4 Remarks. It is crucial to note that x^k cannot be an approximate fixed point of f_k generated by the algorithms of Chapter IV. Indeed, consider the following example where $C = S^2$. Let $F(z) = D$ for all z close to x. Then however fine the triangulation G, there is a p.l. approximation f to F with respect to G so that a simplex containing x is completely-labelled by the standard labelling of Chapter IV. (The diagram shows how such a p.l. approximation f can be

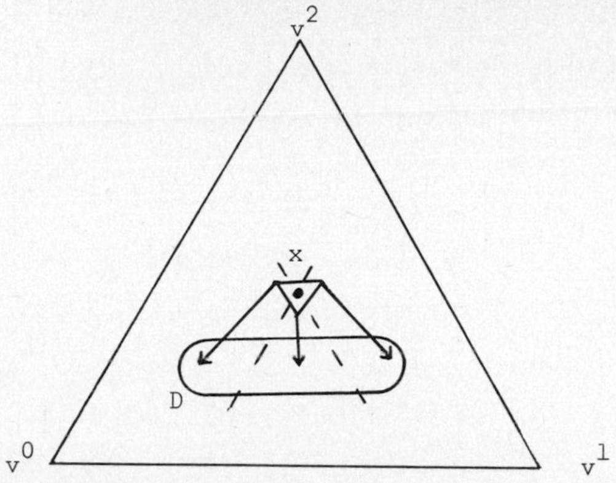

constructed.) However, clearly x is not a fixed point.

It is also important to notice the ease with which the mapping F was extended to F_0, compared to the difficulties using retractions in II.2.

We develop algorithms to find the fixed points of p.l. approximations later. Let us establish how close such points are to their images under the original mapping.

2.5 Theorem

(a) Let $C \subseteq R^n$ be triangulated by G, with $\text{mesh}_2 G \leq \delta$. Let $F: C \to C^*$ satisfy the condition: for all $x \in C$, $y \in B(x,\delta)$, we have $F(y) \subseteq B(F(x),\varepsilon)$. Let f be a p.l. approximation to F w.r.t. G and x^* be a fixed point of f. Then $x^* \in B(F(x^*),\varepsilon)$.

(b) Let $F: C \to C^*$ be as in (a). Pick $c \in \text{int } C$. Let $F_1(x) = F(x)$ if $x \in C$, $\emptyset$ if $x \notin C$, $F_2(x) = \{c\}$ if $x \notin \text{int } C$, $\emptyset$ if $x \in \text{int } C$, and $F_0: R^n \to C^*$ be defined by $F_0 = \text{conv}(F_1 \cup F_2)$. Let R^n be triangulated by G with $\text{mesh}_2 G \leq \delta$, f be a p.l. approximation to F_0 w.r.t. G, and x^* be a fixed point of f. If $B(c,\mu) \subseteq C \subseteq B(c,\nu)$ with $\mu > \varepsilon$, then $x^* \in B(F(x^*),\ \varepsilon + (\delta+\varepsilon)(\nu+\varepsilon)/\mu)$.

(c) Let $C \subseteq R^n$ be triangulated by G with $\text{mesh}_2 G \leq \delta$. Assume $\dim C = n$. Let $f_0: C \to C$ be continuously differentiable with derivative f_0' satisfying $\|f_0'(x) - f_0'(z)\|_2 \leq M\|x - z\|_2$ for all $x,z \in C$. Let f be the p.l. approximation to $\{f_0\}$ w.r.t. G and x^* a fixed point of f. Then $\|f_0(x^*) - x^*\|_2 < \frac{1}{2} M\delta^2$.

Proof.

(a) We have $x^* = \sum_{i \in N_0} \lambda_i y^i = \sum_{i \in N_0} \lambda_i f(y^i)$ with $\lambda_i \geq 0$, $\sum_{i \in N_0} \lambda_i = 1$

and $\langle y^0,\dots,y^n\rangle \in G$. (We assume without loss of generality that $\dim C = n$.) Now $\|y^i - x^*\|_2 \le \delta$, so $f(y^i) \in F(y^i) \subseteq B(F(x^*),\varepsilon)$. Since $B(F(x^*),\varepsilon)$ is convex, it contains all convex combinations of the $f(y^i)$'s, in particular, x^*.

(b) By the argument of (a) we have $x^* = \lambda c + (1-\lambda)d$ for some $0 \le \lambda \le 1$, where $d \in B(F(x^*),\varepsilon)$ and $\lambda = 0$ unless x^* is within δ of $R^n \sim \text{int } C$. If $\lambda = 0$ we are done; hence assume $\lambda > 0$. Because $x^* - d = \lambda(c-d)$, we have

$$x^* \in B(F(x^*),\ \varepsilon + \lambda\|c - d\|_2).$$

Now $F(x^*) \subseteq C$, so $d \in B(C,\varepsilon)$. Let $d' \in C \cap B(d,\varepsilon)$. Since $B(c,\mu) \subseteq C$, $B(\lambda c + (1-\lambda)d',\lambda\mu) \subseteq C$. Also, $\|x^* - (\lambda c - (1-\lambda)d')\|_2 = (1-\lambda)\|d - d'\|_2 \le (1-\lambda)\varepsilon$. Because x^* is within δ of $R^n \sim \text{int } C$, we must have $\delta \ge \lambda\mu - (1-\lambda)\varepsilon \ge \lambda\mu - \varepsilon$. Hence $\lambda \le (\delta+\varepsilon)/\mu$ and $\|c - d\|_2 \le \|c - d'\|_2 + \|d' - d\|_2 \le \nu + \varepsilon$.

(c) For any $x,z \in C$, we have $f_0(z) - f_0(x) = \int_0^1 f_0'(x + \lambda(z-x))(z-x)d\lambda = \int_0^1 [f_0'(x + \lambda(z-x)) - f_0'(x)](z-x)d\lambda + f_0'(x)(z-x)$. Hence

$$\begin{aligned}\|f_0(z) - f_0(x) - f_0'(x)(z-x)\|_2 &= \left\|\int_0^1 [f_0'(x + \lambda(z-x)) - f_0'(x)](z-x)d\lambda\right\|_2 \\ &\le \int_0^1 \|f_0'(x + \lambda(z-x)) - f_0'(x)\|_2\|z - x\|_2 d\lambda \\ &\le \int_0^1 M\lambda\|z - x\|_2^2 d\lambda = \frac{1}{2} M\|z - x\|_2^2.\end{aligned}$$

Now

$$\begin{aligned}x^* = \textstyle\sum_{i\in N_0} \lambda_i y^i &= \textstyle\sum_{i\in N_0} \lambda_i f_0(y^i) \quad \text{with } \lambda_i \ge 0,\ \sum_{i\in N_0} \lambda_i = 1,\ \langle y^0,\dots,y^n\rangle \in G \\ &= \textstyle\sum_{i\in N_0} \lambda_i f_0(x^*) + \sum_{i\in N_0} \lambda_i (f_0(y^i) - f_0(x^*) - f_0'(x^*)(y^i - x^*)) \\ &\qquad + \textstyle\sum_{i\in N_0} \lambda_i f_0'(x^*)(y^i - x^*) \\ &= f_0(x^*) + \textstyle\sum_{i\in N_0} \lambda_i (f_0(y^i) - f_0(x^*) - f_0'(x^*)(y^i - x^*))\end{aligned}$$

Since each summand has norm at most $\frac{1}{2}\lambda_i M\delta^2$, the result follows. □

V.3 Extensions of Kakutani's Theorem

3.1 Corollary (Eaves [9]). Let $C \subseteq R^n$ be compact and convex with $c \in \text{int } C$. Let $F\colon C \to R^{n*}$ be a u.s.c. mapping satisfying $c \in F(x)$ whenever $x \in \partial C$. Then

F has a fixed point in C.

Proof. Since F is u.s.c., $F(C)$ and hence $C' = \mathrm{conv}(C \cup F(C))$ is compact. Extend F to $F_0\colon C' \to C'^*$ using $c \in \mathrm{int}\, C$ in the same way F was extended to F_0 in 2.1. F_0 has a fixed point x^* by 2.1. If $x^* \notin C$, $F_0(x^*) = \{c\} \subseteq C$, a contradiction. Thus $x^* \in C$ and $x^* \in F_0(x^*) = F(x^*)$. □

The next result is useful in the computation of economic equilibria.

3.2 Corollary. Let $T^n = \mathrm{aff}(S^n) = \{x \in R^{n+1} | v^T x = 1\}$ and let $F\colon S^n \to T^{n*}$ be u.s.c. Suppose there exist $f^i \in T^n - T^n$, $i \in N_0$, such that

(i) $x \in S_i^n \Rightarrow x + f^i \in F(x)$; and

(ii) $f_j^i < 0$ for $j \neq i$.

Then F has a fixed point in S^n.

Proof. Let $\tilde{S}^n = \{x \in T^n | x \geq -v\}$. Define u.s.c. mappings F_i, $i = 1,2,3$, as follows. $F_1(x) = F(x)$ if $x \in S^n$, $\emptyset$ if $x \in \tilde{S}^n \sim S^n$; $F_2(x) = \emptyset$ if $x \in \mathrm{rel\ int}\, S^n$, $\{x\} + \mathrm{conv}\{f^i | i \in I\}$ if $x \in \tilde{S}^n \sim \mathrm{int}\, S^n$ and $I = \{i | x_i = \min_j x_j\}$; and $F_3(x) = \emptyset$ if $x \in \mathrm{rel\ int}\, \tilde{S}^n$, $\{(n+1)^{-1} v\}$ if $x \in \partial\tilde{S}^n$. Finally, let $F_0 = \mathrm{conv}(F_1 \cup F_2 \cup F_3)\colon \tilde{S}^n \to T^{n*}$. Then F_0 is u.s.c. and $(n+1)^{-1} v \in F_0(x)$ for $x \in \partial\tilde{S}^n$. By 3.1, F_0 has a fixed point x^* in $\tilde{S}^n$.

If $x^* \notin S^n$, let $I = \{i | x_i^* = \min_k x_k^*\}$. Let z be any member of $F_0(x^*) \subseteq \mathrm{conv}\{(n+1)^{-1} v, \{x^* + f^i | i \in I\}\}$. We show that $\sum_{j\in I} z_j > \sum_{j\in I} x_j^*$ and hence $x^* \neq z$.

If $z = (n+1)^{-1} v$, $\sum_{j\in I} z_j > 0 > \sum_{j\in I} x_j^*$. If $z = x^* + f^i$, $i \in I$, then $\sum_{j\in I} z_j = \sum_{j\in I} (x_j^* + f_j^i) = \sum_{j\in I} x_j^* - \sum_{k\in N_0 \sim I} f_k^i > \sum_{j\in I} x_j^*$ by (ii); note that $I \neq N_0$ since $x^* \neq (n+1)^{-1} v$. Since the inequality holds for $z = (n+1)^{-1} v$ and $z = x^* + f^i$, $i \in I$, it holds for any z in their convex hull $F_0(x^*)$.

We conclude that $x^* \in S^n$ so that $x^* \in F_0(x^*) = F(x^*)$. □

For $n = 2$, a typical choice of f^i's is $f^0 = (1,-\frac{1}{2},-\frac{1}{2})^T$, $f^1 = (-\frac{1}{2},1,-\frac{1}{2})^T$, $f^2 = (-\frac{1}{2},-\frac{1}{2},1)^T$. Then the hypothesis of 3.2 requires that points on the boundary of S^2 are mapped as shown below.

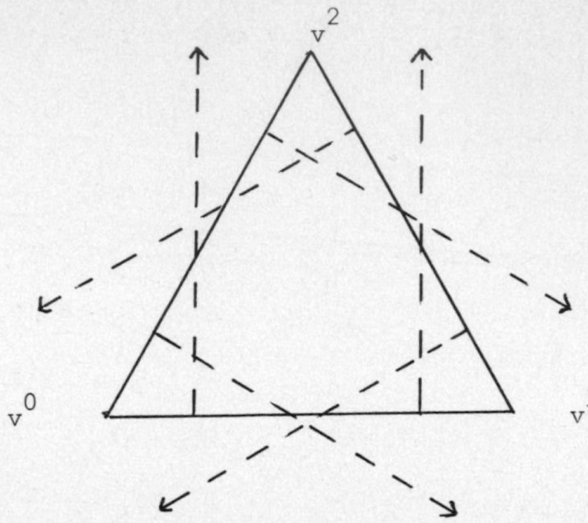

<u>3.3 Corollary</u>. Let $F: R^n \to R^{n*}$ be u.s.c. Suppose there is $x^0 \in R^n$ and $\mu > 0$ such that whenever $x \notin B(x^0,\mu)$ there is some $w \in R^n$ with $w^T(x^0-x) > 0$ and $w^T(f-x) > 0$ for all $f \in F(x)$. Then F has a fixed point in $B(x^0,\mu)$.

<u>Proof</u>. Let C be the compact convex set $B(x^0,2\mu)$. Let F_1 be the restriction of F to C, F_2 be $\emptyset$ on $\operatorname{int} C$ and $\{x^0\}$ on ∂C, and F_0 be $\operatorname{conv}(F_1 \cup F_2)$. Then F_0 is a u.s.c. mapping from C to R^{n*} with $x^0 \in F_0(x)$ whenever $x \in \partial C$. By 3.1, F_0 has a fixed point x^*. If $x^* \in B(x^0,\mu)$, $F_0(x^*) = F(x^*)$ and the result is proved.

Suppose $x^* \notin B(x^0,\mu)$. Then $x^* \in F_0(x^*) \subseteq \operatorname{conv}\{x^0,F(x^*)\}$. Thus $x^* = \lambda x^0 + (1-\lambda)f$ for some $\lambda \in [0,1]$, $f \in F(x^*)$. But then there is $w \in R^n$ with $w^T(x^0-x^*) > 0$ and $w^T(f-x^*) > 0$; hence $w^T(x^*-x^*) > 0$, a contradiction. □

Notice that the result also holds if the condition is required only for $x \in \partial B(x^0,\mu)$. Since the condition is equivalent to $x - x^0 \notin \bigcup_{\lambda \geq 0} \lambda(F(x) - \{x\})$, the result is then an extension of that of I.5.2.

For practical computation we need a slight strengthening of 3.3. The following result is due to Merrill [48,49].

<u>3.4 Corollary</u>. Let $F: R^n \to R^{n*}$ be u.s.c. Suppose there are $x^0 \in R^n$, $\mu > 0$ and $\delta > 0$ such that whenever $x \notin B(x^0,\mu)$, $f \in F(x)$ and $z \in B(x,\delta)$, $(f-x)^T(x^0-z) > 0$. Then F has a fixed point in $B(x^0,\mu)$.

<u>Proof</u>. The condition implies that $w = x^0 - z$ satisfies the hypotheses of 3.3. □

Again we can strengthen 3.4 by requiring the condition to hold only when $x \in \partial B(x^0,\mu)$, but in this case the result is unsuitable for computation. If the condition need only hold for $x \in B(x^0,\mu+2\delta) \sim B(x^0,\mu)$, then a result is obtained which is appropriate for computation; on the other hand it is not clear whether the applicability of the result is appreciably extended for practical problems.

We will develop algorithms based on 3.1 and 3.4. For those based on 3.1, we do not need C explicitly. We must have available a half space containing C and $c \in \text{int } C$, and, for any $x \in R^n$, we must be able to determine whether or not x lies in C. The algorithms based on 3.4 require only a knowledge of δ. Fortunately, for many problems the condition of 3.5 holds for any $\delta > 0$ with appropriate x^0 and μ. Clearly, if $F(R^n)$ is compact (as it is, for example, if F is the natural extension to R^n of a mapping satisfying the conditions of 3.1), then this stronger condition holds. For all algorithms, as indicated by the proof of 2.1, we need only find a single member of $F(x)$ for each $x \in R^n$.

CHAPTER VI: APPLICATIONS OF KAKUTANI'S THEOREM AND ITS EXTENSIONS

We now show how the ability to find approximate fixed points of upper semi-continuous mappings enables one to solve a wide range of important problems, without the limitations of Chapter II. Again we point out that constructing an appropriate mapping is relatively straightforward. Indeed, for any given problem it is easier than constructing a continuous function, since u.s.c. mappings can be patched together so easily.

VI.1 Equilibria in Finite Economies with Production. Consider the model of an exchange economy as in II.1. The demand side of the economy we will consider is the same except for the following important relaxation. For any $p \in S^n$, we assume that the ith consumer has a corresponding set $D^i(p)$ of possible demands. D^i is assumed to be a u.s.c. mapping from S^n to R_+^{n+1*} (see exercise 5.1); set $D(p) = \sum_{i=1}^m D^i(p)$. The aggregate demand correspondence $D: S^n \to R_+^{n+1*}$ is then u.s.c.; assume that it satisfies Walras' Law: for all $p \in S^n$, $d \in D(p)$, $p^T d = p^T w$.

On the production side we restrict ourselves to a simple model. Let $B = [-I, C]$ be an $(n+1)\times(k+1)$ matrix. For $0 \le j \le k$, the jth column B_j of B represents an activity; if activity j is used at level 1, $|b_{ij}|$ units of commodity i are supplied as output (if $b_{ij} \ge 0$) or required as input (if $b_{ij} < 0$). Note that the first $n+1$ columns of B correspond to an assumption of free disposal.

The set of all production vectors is $\{Bz | z \ge 0\}$. The model can be generalized to allow more general production sets, but they must at least be convex. We make the reasonable assumption that $\{Bz|z \ge 0\} \cap R_+^{n+1} = \{0\}$; production is bounded. Equivalently, one assumes that there is $q > 0$ with $qB < 0$.

We can now define an equilibrium as a price vector and a vector of activity levels, such that the sum of the initial resources and the production vector is a member of the demand set for those prices. (Since free disposal is included in the production possibilities, we need not worry about excess supply.) By permitting only activities that maximize profit we ensure that producers have an incentive to use them.

1.1 Definition. An equilibrium is a pair $(p^*,z^*) \in S^n \times R_+^{k+1}$ such that $w + Bz^* \in D(p^*)$ and $p^{*T}B \leq 0$.

Using Walras' Law we find that $p^{*T}w + p^{*T}Bz^* = p^{*T}w$, and hence $p^{*T}Bz^* = 0$, $p^{*T}B \leq 0$. The interpretation is that no activity makes a positive profit; those activities used (at positive level) must make zero profit. In particular, if a free disposal activity is used, the price of the corresponding commodity is zero.

We shall prove that an equilibrium exists (if $w > 0$) by constructing a u.s.c. mapping satisfying the hypotheses of V.3.2. Scarf [58] does not require $w > 0$ and does not need the vector q above, as we do. However, our mapping allows the use of more sophisticated algorithms for fixed-point problems.

1.2 Theorem. If $w > 0$ then an equilibrium exists.

Proof. Let $p \in S^n$ be given. If p is not an equilibrium price vector, we wish to modify it suitably. First, if any $p_i = 0$, we may want to increase p_i to avoid leaving S^n. Secondly, if $p > 0$ but $\pi^* = \max_r p^T B_r > 0$, we may want to adjust p to decrease the profit of the most profitable activities. Finally, if $p > 0$ and $\pi^* < 0$, we adjust p to decrease demand.

Define the following mappings:

$$E_1(p) = \begin{cases} \operatorname{conv}\{v^i \mid p_i = 0\} & \text{if } p \in \partial S^n; \\ \emptyset & \text{otherwise.} \end{cases}$$

$$E_2(p) = \begin{cases} \operatorname{conv}\{-B_r \mid p^T B_r = \pi^* = \max_s p^T B_s\} & \text{if } \pi^* \geq 0; \\ \emptyset & \text{otherwise.} \end{cases}$$

$$E_3(p) = \begin{cases} D(p) & \text{if } \pi^* \leq 0; \\ \emptyset & \text{otherwise.} \end{cases}$$

Clearly, E_i is u.s.c. for $i = 1,2,3$; for example, E_2 is the convex hull of mappings such as $E_2^r(p) = \begin{cases} -B_r & \text{if } p^T B_r \geq 0,\ p^T(B_r - B_s) \geq 0 \\ \emptyset & \text{otherwise} \end{cases}$ which are u.s.c. by

V.1.4(e). Now let $E = \text{conv}(E_1 \cup E_2 \cup E_3)$. Then $E: S^n \to R^{n+1*}$ is u.s.c. We would like to send p to $\{p\} + E(p)$, but unfortunately this set may not even lie in $T^n = \{x \in R^{n+1} | v^T x = 1\}$. We must modify $E(p)$ so that $e \in E'(p) \Rightarrow e \in T^n - T^n$. For any $e \in E(p)$, proceed as follows.

Let $e'(e)$ have coordinates e_i', $i \in N_0$, defined by

$$e_i' = q_i e_i (q^T w / q^T e) - q_i w_i.$$

To show that e' is well-defined and $e'(\cdot)$ is a continuous function, we need only show that $q^T e$ is bounded away from zero. But for $e = v^i$, $q^T e = q_i > 0$. For $e = -B_r$, $q^T e = -q^T B_r \geq \min_s(-q^T B_s) > 0$. For $e = d \in D(p)$, $q^T e \geq (\min_i q_i) v^T e \geq (\min_i q_i) p^T e = (\min_i q_i) p^T w \geq (\min_i q_i)(\min_j w_j) > 0$. Since $E(p) = \text{conv}(E_1(p) \cup E_2(p) \cup E_3(p))$, we have $q^T e \geq \min\{\min_i q_i, \min_s(-q^T B_s), (\min_i q_i) \times (\min_j w_j)\} > 0$.

Thus we may set $E'(p) = \{e'(e) | e \in E(p)\}$. By V.1.4(b), E' is u.s.c. Note that $e'(e)$ lies in $T^n - T^n$ for any e with $q^T e \neq 0$; it is an easy exercise (5.2) to show that $E'(p)$ is convex. Thus $E': S^n \to (T^n - T^n)^*$, and $F: S^n \to T^{n*}$ is u.s.c. with $F(p) = \{p\} + E'(p)$.

Now if $p \in S_i^n$, $v^i \in E(p)$, so $e'(v^i) \in E'(p)$. For each $i \in N_0$ define $f^i = e'(v^i) = (-q_0 w_0, \ldots, -q_i w_i + q^T w, \ldots, -q_n w_n)^T$. Thus for $p \in S_i^n$, $p + f^i \in F(p)$. Thus the hypotheses of V.3.2 are satisfied, and F has a fixed point $p^* \in S^n$.

Then $0 \in E'(p^*)$, that is, there is $e \in E(p^*)$ with $e = \lambda w$ for some positive scalar λ. From the definition of E we have

$$\lambda w = \textstyle\sum_{i \in N_0} \mu_i v^i + \sum_{r=0}^{k} \nu_r (-B_r) + \rho d \qquad (*)$$

with $\mu_i, \nu_r, \rho \geq 0$, $\sum \mu_i + \sum \nu_r + \rho = 1$. In addition, $\mu_i = 0$ unless $p_i^* = 0$, $\nu_r = 0$ unless $p^{*T} B_r = \max_s p^{*T} B_s = \pi^* \geq 0$, $\rho = 0$ unless $\pi^* \leq 0$, and $d \in D(p^*)$.

Premultiplying (*) by p^{*T} yields

$$\lambda p^{*T} w = 0 + (\textstyle\sum - \nu_r)\pi^* + \rho p^{*T} d. \qquad (**)$$

If $\pi^* > 0$, $\rho = 0$. Then the left-hand side of (**) is positive and the right-hand side nonpositive, a contradiction.

If $\pi^* < 0$, all $\nu_r = 0$, and (**) becomes $\lambda p^{*T}w = \rho p^{*T}d$. Using Walras' Law, we obtain $\lambda = \rho$. Dividing (*) by λ gives

$$w + \sum_{i \in N_0} (\mu_i/\lambda)(-v^i) = d.$$

Hence $(p^*, (\mu_0/\lambda,\ldots,\mu_n/\lambda,0,\ldots,0)^T)$ is an equilibrium; the only activities possibly used are those of free disposal.

Finally, if $\pi^* = 0$, then again $\lambda p^{*T}w = \rho p^{*T}d$ and $\lambda = \rho$. Since $v^i = -B_i$, $i \in N_0$, and $p_i^* = 0$ implies that activity B_i makes the maximum profit zero, we can incorporate all the μ_i's that are non-zero in the ν_r's. Then from (*) we have

$$w + \sum_{r=0}^{k} (\nu_r/\lambda)B_r = d.$$

Hence $(p^*, (\nu_0/\lambda,\ldots,\nu_k/\lambda)^T)$ is an equilibrium and the theorem is proved. □

VI.2 Unconstrained Optimization. The next three sections follow Merrill's analysis [48,49] closely. We restrict ourselves to the simplest case.

Let $\theta: R^n \to R$ be convex. We write $\text{lev}_\alpha\theta$ for $\{x \in R^n | \theta(x) \leq \alpha\}$ and $D\theta$ for the subdifferential mapping of V.1. Then with θ finite everywhere, we conclude that θ is continuous, $\text{lev}_\alpha\theta$ is closed and convex, and $D\theta: R^n \to R^{n*}$ is u.s.c. (Rockafellar [51], theorems 10.1.1, 4.6, 23.4 and 24.5.1). Also, if $\text{lev}_\alpha\theta$ is nonempty and bounded for some α, it is bounded for all α (corollary 8.7.1 of Rockafellar [51]).

Define $F_1: R^n \to R^{n*}$ by $F_1(x) = \{x\} - D\theta(x)$. Let $\theta(c) < \alpha$, and define $F'(x)$ to be $F_1(x)$ on $\text{lev}_\alpha\theta$, $\emptyset$ otherwise, $F''(x)$ to be $\{c\}$ on $R^n \sim \text{int lev}_\alpha\theta$, $\emptyset$ otherwise, and $F_2 = \text{conv}(F' \cup F'')$.

2.1 Lemma. The set of fixed points of F_i is the set of minimizers of θ, for $i = 1,2$.

Proof. The result is clear for F_1. For F_2 we need only verify that F_2 has no fixed points on $\partial\ \mathrm{lev}_\alpha \theta$. But if $x \in \partial\ \mathrm{lev}_\alpha\theta$, $x \in F_2(x)$, then $x = \lambda c + (1-\lambda)(x-d)$, for $\lambda \in [0,1]$ and $d \in D\theta(x)$. Then $d = \lambda(1-\lambda)^{-1}(c-x)$ and $\theta(c) > \theta(x)$, a contradiction. □

2.2 Theorem. Suppose that θ has a bounded nonempty level set. Then F_2 restricted to $C = \mathrm{lev}_\alpha\theta$ satisfies the hypothesis of V.3.1 and F_1 satisfies the hypothesis of V.3.4. Indeed, for any $x^0 \in R^n$ and $\delta > 0$ there is $\mu > 0$ satisfying the condition of V.3.4.

Proof. Since C is compact, the first part is clear. Now let $\beta = \sup\{\theta(x) | x \in B(x^0,\delta)\} < \infty$. Then $\mathrm{lev}_\beta\theta$ is bounded; choose μ so that $\mathrm{lev}_\beta\theta \subseteq B(x^0,\mu)$. Let $x \notin B(x^0,\mu)$, $f \in F_1(x)$ and $z \in B(x,\delta)$. Then $x - f \in D\theta(x)$ and hence $(x-f)^T(x^0-z) = (x-f)^T((x^0-z+x) - x) \leq \theta(x^0-z+x) - \theta(x)$. But $x^0 - z + x \in B(x^0,\delta)$; thus $(x-f)^T(x^0-z) < \beta - \beta = 0$. □

If a fixed point of a p.l. approximation to F_1 is found, a lower bound on θ is naturally generated. If x^* is such a point, $x^* = \sum_{i\in N_0} \lambda_i y^i = \sum_{i\in N_0} \lambda_i(y^i-d^i)$, where $\lambda_i \geq 0$, $\sum_{i\in N_0} \lambda_i = 1$ and $d^i \in D\theta(y^i)$ for each $i \in N_0$. For every $x \in R^n$ and each $i \in N_0$, $\theta(x) \geq \theta(y^i) + d^{iT}(x-y^i)$; hence

$$\theta(x) \geq \sum_{i\in N_0} \lambda_i\theta(y^i) + \sum_{i\in N_0} \lambda_i d^{iT}(x-y^i)$$

$$= \sum_{i\in N_0} \lambda_i\theta(y^i) + \sum_{i\in N_0} \lambda_i d^{iT}(x^*-y^i) \qquad (*)$$

since $\sum \lambda_i d^i = 0$. If $\mathrm{diam}\langle y^0,\ldots,y^n\rangle$ is small, the last term in (*) is likely to be small; the first term is at least $\theta(x^*)$, showing x^* to be close to minimizing θ.

Merrill [48,49] discusses extensions of this application, including the case where θ is not convex.

VI.3 Constrained Optimization.

Consider the problem: minimize $\theta(x)$ subject to $g_k(x) \leq 0$, $k = 1,2,\ldots,m$, where θ and each g_k is convex and takes R^n to R.

By defining $\psi(x)$ to be $\max_k g_k(x)$, we reduce this problem to

$$\min\{\theta(x)|\psi(x) \leq 0\}, \qquad (P)$$

where θ and ψ are convex. If each g_k is differentiable, then $D\psi(x) = \text{conv}\{\nabla g_k(x)|g_k(x) = \psi(x)\}$.

A natural modification of any $x \in R^n$ that does not solve (P) motivates the following definitions. Let

$$H_1(x) = \begin{cases} \{x\} - D\theta(x) & \text{if } x \in \text{lev}_0\psi, \\ \emptyset & \text{otherwise;} \end{cases}$$

$$H_2(x) = \begin{cases} \emptyset & \text{if } x \in \text{int lev}_0\psi, \\ \{x\} - D\psi(x) & \text{otherwise;} \end{cases}$$

and $F_1 = \text{conv}(H_1 \cup H_2): R^n \to R^{n*}$.

If there is a $c \in R^n$ with $\psi(c) < 0$, we define F_2 as follows. Let $\theta(c) < \alpha$, $\psi(c) < 0 < \beta$, and $C = \text{lev}_\alpha\theta \cap \text{lev}_\beta\psi$. Define $F'(x)$ to be $F_1(x)$ if $x \in C$ and $F''(x)$ to be $\emptyset$ if $x \in \text{int } C$, $\{c\}$ if $x \in \partial C$. Finally, let $F_2 = \text{conv}(F' \cup F''): C \to R^{n*}$.

We write $\inf \psi$ to denote $\inf\{\psi(x)|x \in R^n\}$. The relationship between F_1, F_2, and (P) is shown by

3.1 Lemma. Any fixed point of F_1 lies in $\text{lev}_0\psi$ if this set is nonempty. If $\inf \psi < 0$, then the set of optimal solutions to (P) is the set of fixed points of F_i, $i = 1$ or 2.

Proof. Let $x^* \in F_1(x^*) \sim \text{lev}_0\psi$. Then $0 \in D\psi(x^*)$ and ψ takes on its minimum at x^*. Thus $\inf \psi = \psi(x^*) > 0$ and $\text{lev}_0\psi$ is empty.

Now assume $\inf \psi < 0$. Let x^* be an optimal solution to (P). It follows (see Rockafellar [51], corollary 28.2.1) that x^* minimizes $\theta + \lambda\psi$ for some $\lambda \geq 0$ with $\lambda = 0$ unless $\psi(x^*) = 0$. Thus $0 = (1+\lambda)^{-1}d + \lambda(1+\lambda)^{-1}e$ with

$d \in D\theta(x^*)$, $e \in D\psi(x^*)$, and $x^* \in F_1(x^*)$. Since $\theta(x^*) \leq \theta(c) < \alpha$ and $\psi(x^*) \leq 0 < \beta$, $x^* \in \text{int } C$ and $x^* \in F_2(x^*)$ also.

Conversely, let x^* be a fixed point of F_1. If $\psi(x^*) < 0$, $0 \in D\theta(x^*)$ and hence x^* solves (P). If $\psi(x^*) = 0$, $0 = \lambda d + (1-\lambda)e$ for some $\lambda \in [0,1]$, $d \in D\theta(x^*)$ and $e \in D\psi(x^*)$. If $\lambda = 0$, x^* minimizes ψ, a contradiction. Dividing by $\lambda > 0$ we obtain $d + \lambda^{-1}(1-\lambda)e = 0$, and x^* minimizes $\theta + \lambda^{-1}(1-\lambda)\psi$. Thus x^* solves (P).

Let x^* be a fixed point of F_2. Clearly, $x^* \in C$, and if $x^* \in \text{int } C$, x^* is a fixed point of F_1 and hence an optimal solution to (P). Let $x^* \in \partial C$. If $x^* \notin \text{lev}_0\psi$, we obtain a contradiction as in 2.1. Thus $\psi(x^*) \leq 0 < \beta$ and $\theta(x^*) = \alpha$. If $\psi(x^*) < 0$, then $0 \in \text{conv}(D\theta(x^*) \cup \{x-c\})$ and a contradiction arises as in 2.1. If $\psi(x^*) = 0$, then $0 \in \text{conv}(D\chi(x^*) \cup \{x-c\})$ with $\chi = \lambda\theta + (1-\lambda)\psi$, some $\lambda \in [0,1]$; again a contradiction results from $\theta(x^*) = \alpha > \theta(c)$, $\psi(x^*) = 0 > \psi(c)$ and hence $\chi(x^*) > \chi(c)$. □

If $\text{lev}_\alpha\theta \cap \text{lev}_\beta\psi$ is nonempty and bounded for some α, β, it is bounded for all α, β. (Probably the easiest way to see this is by considering functions of the form $\phi(x) = \max\{\lambda(\theta(x) - \alpha), \mu(\psi(x) - \beta)\}$, and applying corollary 8.7.1 of Rockafellar [51].) This condition is sufficient for the existence of fixed points of F_1 and F_2. More precisely, we have

<u>3.2 Theorem</u>. If ψ has a nonempty bounded level set, then for any $x^0 \in R^n$, $\delta > 0$ there is a $\mu > 0$ satisfying V.3.4 for F_1. If $\inf \psi < 0$ and $\text{lev}_\alpha\theta \cap \text{lev}_\beta\psi$ is nonempty and bounded for some α, β, then F_2 satisfies the conditions of V.3.1, and there are $x^0 \in R^n$, $\delta > 0$, and $\mu > 0$ satisfying V.3.4 for F_1.

<u>Proof</u>. If ψ has a nonempty bounded level set, then $\text{lev}_0\psi$ is bounded and $F_1(x) = H_2(x)$ for x sufficiently far from 0. The first conclusion then follows as in 2.2. The result for F_2 is clear. If $\inf \psi < 0$, let $\psi(x^0) < 0$ and δ be such that $\beta = \sup\{\psi(x) | x \in B(x^0,\delta)\} < 0$ and let $\alpha = \sup\{\theta(x) | x \in B(x^0,\delta)\}$. Choose μ so that $\text{lev}_\alpha\theta \cap \text{lev}_\beta\psi \subseteq B(x^0,\mu)$. The conclusion follows as in 2.2. □

For computation, the first part of 3.2 is much more satisfactory than the last, for we choose δ arbitrarily. If it is known that there is a feasible solution to $\theta(x) < \alpha$, $\psi(x) < 0$, then (P) is equivalent to: minimize $\theta(x)$ subject to $\psi(x) \leq 0$, $\theta(x) \leq \alpha$, or minimize $\theta(x)$ subject to $\psi'(x) \leq 0$ (P'), where $\psi'(x) = \max\{\psi(x), \theta(x) - \alpha\}$. If $\mathrm{lev}_{\alpha'}\theta \cap \mathrm{lev}_{\beta'}\psi$ is bounded and nonempty for some α', β', then ψ' has a bounded nonempty level set. Thus basing F_1 on (P') rather than (P) allows one to choose δ arbitrarily.

VI.4 The Nonlinear Complementarity Problem

Recall that the NLCP associated with a continuous function $g: R^n_+ \to R^n$ is to find $x^* \geq 0$ with $g(x^*) \geq 0$ and $x^{*T}g(x^*) = 0$. We describe how Merrill [48,49] converts the NLCP into a fixed-point problem and give sufficient conditions for the resultant mapping to satisfy the hypothesis of V.3.4.

Let $\psi: R^n \to R$ be the convex function with $\psi(x) = \max_{i \in N}(-x_i)$; note that $\mathrm{lev}_0\psi = R^n_+$. Define $F_1(x)$ to be $\emptyset$ if $x \in \mathrm{int}\, R^n_+$ and $\{x\} - D\psi(x)$ otherwise, and $F_2(x)$ to be $\{x - g(x)\}$ if $x \in R^n_+$ and $\emptyset$ otherwise. Let $F = \mathrm{conv}(F_1 \cup F_2)$: $R^n \to R^{n*}$.

<u>4.1 Lemma</u>. x is a fixed point of F iff x solves the NLCP associated with g.

<u>Proof</u>. Let x be a fixed point of F. If $x \notin R^n_+$, $0 \in D\psi(x)$ and x minimizes ψ, a contradiction. If $x \in \mathrm{int}\, R^n_+$, $g(x) = 0$ and x solves the NLCP. If $x \in \partial R^n_+$, set $I = \{i \in N | x_i = 0\}$. Then $D\psi(x) = \mathrm{conv}\{-u^i | i \in I\}$ and hence $g_i(x) = 0$ for $i \notin I$ and $g_i(x) \geq 0$ for $i \in I$. Thus x solves the NLCP.

Conversely, given a solution x to the NLCP, set $I = \{i | x_i = 0\}$. If $I = \emptyset$, $g(x) = 0$ and $x \in F_2(x) = F(x)$. Otherwise, $-g(x) \in \mathrm{conv}\{-u^i | i \in I\} = D\psi(x)$ and $x = \frac{1}{2}(x + g(x)) + \frac{1}{2}(x - g(x)) \in \mathrm{conv}(F_1(x) \cup F_2(x)) = F(x)$. □

A sufficient condition for F to satisfy the hypothesis of V.3.4 (and thus for the NLCP to have a solution) is given by

<u>4.2 Theorem</u>. Suppose there exist $\alpha > 0$ and $\nu > 0$ such that whenever $x \in R^n_+ \sim B(0,\nu)$, $x^T g(x) > \alpha \|x\|_2 \|g(x)\|_2$. Then for any $\delta > 0$ there exist $x^0 \in R^n$

and $\mu > 0$ satisfying the hypothesis of V.3.4 for F.

<u>Proof</u>. Choose $x^0 > \delta u$ and $\mu > \max\{\nu + \|x^0\|_2, (\delta + \|x^0\|_2)/\alpha\}$. Note that $\psi(x) < 0$ for any $x \in B(x^0,\delta)$. Now let $x \notin B(x^0,\mu)$, $f \in F(x)$, and $z \in B(x,\delta)$.

If $\psi(x) > 0$, then $-(f-x) \in D\psi(x)$. Thus, as in 2.2, we find $(f-x)^T(x^0-z) \geq \psi(x) - \psi(x^0+x-z) > 0$.

If $\psi(x) < 0$, $x \in R^n_+$ and $x \notin B(0,\nu)$. In this case $f = x - g(x)$ and $(f-x)^T(x^0-z) = -g(x)^T(x^0-z) = g(x)^T x + g(x)^T(-x^0-x+z) > \alpha\|x\|_2\|g(x)\|_2 - \|g(x)\|_2\|-x^0-x+z\| \geq \|g(x)\|_2(\alpha\|x\|_2 - \|x^0\|_2 - \delta) > 0$, since $g(x)$ cannot be 0.

Finally, if $\psi(x) = 0$, a combination of the cases above yields $(f-x)^T(x^0-z) > 0$. □

VI.5 Exercises.

<u>5.1.</u> Under the assumptions of II.4.1, show that D is a u.s.c. mapping on $T^n = \{p \in S^n | p > 0\}$.

<u>5.2.</u> Show that $E'(p)$ in the proof of 1.2 is convex.

<u>5.3.</u> Let $\theta,\psi\colon R^n \to R$ be convex with $\text{lev}_{\alpha'}\theta \cap \text{lev}_{\beta'}\psi$ nonempty and bounded for some α', β'. Show that $\text{lev}_\alpha\theta \cap \text{lev}_\beta\psi$ is bounded for all α, β.

<u>5.4.</u> Let $\theta\colon R^n \to R$ be convex with a bounded nonempty level set. Pick $x^1,x^2 \in R^n$ with $\theta(x^1) \neq \theta(x^2)$. Using only x^i, $\theta(x^i)$, and some $d^i \in D\theta(x^i)$ for $i = 1,2$, show how to obtain: α so that $C = \text{lev}_\alpha\theta$ is compact; some $c \in \text{int } C$; and a halfspace containing C.

<u>5.5.</u> Let $\theta,\psi\colon R^n \to R$ be convex with $\text{lev}_{\alpha'}\theta \cap \text{lev}_{\beta'}\psi$ nonempty and bounded for some α', β'. Pick $x^1,x^2 \in R^n$ with $\psi(x^1) < 0$ and $\theta(x^1) \neq \theta(x^2)$ if $\psi(x^2) \leq 0$. Using only x^i, $\theta(x^i)$, $\psi(x^i)$, some $d^i \in D\theta(x^i)$, and some $e \in D\psi(x^i)$, $i = 1,2$, show how to obtain: $\beta > 0$ and α so that $C = \text{lev}_\alpha\theta \cap \text{lev}_\beta\psi$ is compact; some $c \in \text{int } C$; and a halfspace containing C.

<u>5.6.</u> Suppose the condition of theorem II.3.1 holds for some $w > 0$. Use V.3.1 to prove that the NLCP associated with g has a solution in C_ρ. (Hint: $w \in \text{int } C_\rho$.)

CHAPTER VII: EAVES' FIRST ALGORITHM

The first algorithms for u.s.c. mappings were developed by Hansen and Scarf [29]. However, it is more natural for us to use triangulations rather than primitive sets, and hence we will describe Eaves' algorithm [10,12]. Section 1 introduces the crucial notions of vector labelling and complete simplices. Section 2 describes how the problem to be solved, based on V.3.1, is set up. In Section 3 we define a graph Γ similar to the graph Γ_n of Chapter III. Eaves' algorithm is presented in Section 4.

VII.1 Vector Labelling and Complete Simplices. Let A be an affine subspace in R^m, and suppose $\dim A = n$. Let $F: A \to A^*$ be u.s.c., G be a triangulation of A, and f be a piecewise-linear approximation to F with respect to G. Let S be the linear subspace $A-A$ of dimension n. Define $\ell: A \to S$ by $\ell(x) = f(x) - x$. We call ℓ the vector labelling for f (or a vector labelling for F with respect to G). Often, ℓ is taken to be a function on the vertices of G; however, extending ℓ to R^n shows that fixed points of f are zeroes of ℓ.

If $\sigma = \langle y^0,\ldots,y^k\rangle \in G^k$, then the $(m+1)\times(k+1)$ matrix

$$L_\sigma = \begin{bmatrix} 1 & \ldots & 1 \\ \ell(y^0) & \ldots & \ell(y^k) \end{bmatrix}$$ is called the label matrix of σ.

Let t^0 be the unit vector $(1,0,\ldots,0)^T$ in R^{m+1}. Then we can identify fixed points of ℓ by means of

1.1 Lemma. Let A, F, G, f, and ℓ be as above, and let $\sigma = \langle y^0,\ldots,y^n\rangle \in G$. Then there is a fixed point of f in $\overline{\sigma}$ if and only if there is a solution to

$$L_\sigma w = t^0, \quad w \in R^{n+1}_+. \qquad (*)$$

If w solves (*), then $x^* = \sum_{i\in N_0} w_i y^i$ is a fixed point of f.

Proof. For any $x = \sum_{i\in N_0} w_i y^i \in \overline{\sigma}$ we have

$$\sum_{i\in N_0} w_i \ell(y^i) = \ell(x) = f(x) - x \qquad (\dagger)$$

If x^* is a fixed point of f in $\overline{\sigma}$, then we can find $w \in S^n$ with $x^* = \sum_{i \in N_0} w_i y^i$; and by (†) w is a solution to (*).

Now let w solve (*) and set $x^* = \sum_{i \in N_0} w_i y^i$. Then $w \geq 0$ and $v^T w = 1$, from the first equation in (*); thus $x^* \in \overline{\sigma}$. Now from (†) we find $f(x^*) = x^*$. □

Motivated by 1.1, we state

1.2 Definition. A simplex σ of G is complete if there is a solution to (*).

Vector labelling can be regarded as an extension of integer labelling, and complete simplices of completely-labelled simplices, as follows. Let $g: S^n \to S^n$ be continuous. Suppose we use the integer label $\min\{i | g_i(y) - y_i = \min_k (g_k(y) - y_k)\}$ for y. This labelling is admissible if g has no fixed points on ∂S^n. Now let $A = \text{aff}(S^n)$. For $x \in \text{rel int } S^n$, let $I(x) = \{i | g_i(x) - x_i = \min_k (g_k(x) - x_k)\}$. For $x \in \partial S^n$, let $I(x) = \{i | g_i(x) - x_i = \min_k (g_k(x) - x_k)\} \cup \{i | x_i = \max_k x_k\}$; and for $x \in A \sim S^n$, let $I(x) = \{i | x_i = \max_k x_k\}$.

For each $i \in N_0$ let r^i be the ith column of the $(n+1)\times(n+1)$ matrix

$$\begin{bmatrix} -1 & 0 & \cdots & 0 & +1 \\ +1 & \ddots & & & 0 \\ 0 & \ddots & \ddots & & \vdots \\ \vdots & & \ddots & \ddots & \vdots \\ \vdots & & & -1 & 0 \\ 0 & \cdots & 0 & +1 & -1 \end{bmatrix}$$

. Define $F: A \to A^*$ by $F(x) = \{x\} + \text{conv}\{r^i | i \in I(x)\}$.

For each $i \in N_0$, $\{x \in A | i \in I(x)\}$ is closed; hence F is u.s.c. It is easy to check that the fixed points of F are exactly those of g.

Now let G be a triangulation of A. For $y \in G^0$ choose $f(y) = y + r^i$ with i the least index in $I(y)$, and let f be the corresponding p.l. approximation to F and ℓ its vector labelling. Then if $\sigma \in G$ lies in S^n, it is completely-labelled (with the integer labelling) iff it is complete (with the vector labelling ℓ).

The preceding discussion is based on Eaves [14].

VII.2 Preliminaries. For simplicity we take the set A of Section 1 to have full dimension, i.e., $m = n$. The necessary changes for $m > n$ are straightforward.

Suppose we have a compact convex set $C \subseteq R^n$ and $c \in \text{int } C$. A u.s.c. mapping $F_0: C \to R^{n*}$ is given, with $c \in F_0(x)$ whenever $x \in \partial C$. We must have available c and some halfspace containing C; for simplicity assume $C \subseteq \{x \in R^n | x_1 > 0\}$.

We now extend F_0 to R^n by setting $F_1(x)$ to be $F_0(x)$ if $x \in C$, $\emptyset$ otherwise; $F_2(x)$ to be $\emptyset$ if $x \in C$, $\{c\}$ otherwise; and $F = \text{conv}(F_1 \cup F_2)$: $R^n \to R^{n*}$. Then F is u.s.c. and has the same fixed points as F_0. We will find fixed points of piecewise linear approximations f to F with respect to G. Let $G = \delta K_1$ or δJ_1.

Pick a starting point x^0 with $x_1^0 = 0$. To enhance convergence, x^0 should be close to a fixed point of F--but note that $x^0 \notin C$. In the next chapter we shall see how our starting point can be chosen wherever we wish. If necessary, perturb x^0 so that it lies in some (n-1)-simplex of G^{n-1}. Define $d = x^0 - c$.

Let f be a p.l. approximation to F with respect to G, and ℓ its vector labelling. A fixed point of f can be identified if we can find a complete simplex $\sigma \in G$.

To find c.l. simplices in Chapter III, we had to relax our requirements and search a.c.l. simplices. A similar relaxation is necessary here. Note that L_σ is square; hence the solution to $L_\sigma w = v^0$ will possibly be unique--we do not have the required degree of freedom to produce paths.

To see how to define almost-complete simplices, consider the analogy with integer labelling as in the last section. For the vector labelling there, a simplex σ is a.c.l. iff there is a solution to $L_\sigma w + \binom{1}{r^n}\lambda = t^0$, $w \in R_+^{n+1}$, $\lambda \geq 0$. Thus it is natural to define almost-complete simplices by including an artificial column.

<u>2.1 Definition</u>. Let $\sigma = \langle y^0,\ldots,y^k\rangle \in G^{n-1} \cup G^n$. We call σ <u>almost-complete</u> if there is a solution to

$$L_\sigma w + \binom{1}{d}\lambda = v^0, \quad w \geq 0, \quad \lambda \geq 0. \qquad (*)$$

When $\sigma \in G^n$ the linear system above has $n+1$ rows and $n+2$ columns. If the matrix $[L_\sigma, \binom{1}{d}]$ has rank $n+1$, barring degeneracy, there will be two solutions to (*) with $n+1$ variables positive. We would like this to be true in general, but we cannot assume that degeneracy does not occur. In most linear programming problems this assumption is fairly harmless; if degeneracy occurs, the objective function usually acts to avoid cycling. In fixed-point algorithms, however, the columns of L_σ are likely to be close if the grid size δ is small; with roundoff error, degeneracy can easily occur. Since convergence depends critically on unique pivoting, we must devise a method of circumventing degeneracy.

We do so by using lexicographic systems. A row vector e is lexicographically positive if $e \neq 0$ and $e_j > 0$, where $j = \min\{i | e_i \neq 0\}$; we write $e \succ 0$. We say e is lexicographically nonnegative and write $e \succeq 0$ if $e \succ 0$ or $e = 0$. A matrix is lexicographically positive (nonnegative) if each of its rows is.

Let A and B be real matrices of dimensions $m\times(m+1)$ and $m\times p$, and suppose A and B have rank m. Then any solution X to the linear system $AX = B$, $X \succeq 0$, has rank m; hence X has at most one zero row. If X has one zero row, the columns of A corresponding to the remaining rows of X are linearly independent and form a basis for R^m. In this case we call X a basic feasible solution to $AX = B$, $X \succeq 0$.

Noting that the null space of A has dimension 1 one can easily prove the following standard result of linear programming.

<u>2.2 Theorem</u>. Let A and B be as above. If $AX = B$, $X \succeq 0$ has a solution, it has exactly two basic feasible solutions. $\square$

See Dantzig [8] for a further discussion of lexicographic rules. Note that one basic feasible solution can be obtained from the other by a linear programming pivot step.

A natural way to obtain a right-hand side of rank $n+1$ in our systems is to replace v^0 by the identity matrix I of order $n+1$.

<u>2.3 Definition</u>. Let $\sigma = \langle y^0,\ldots,y^k\rangle \in G^{n-1} \cup G^n$. We call σ very complete (v.c.) if there is a solution to

$$L_\sigma W = I, \quad W \succeq 0, \tag{i}$$

and very almost complete (v.a.c.) if there is a solution to

$$L_\sigma W + (\tfrac{1}{d})e = I, \quad W \geq 0, \quad e \geq 0. \qquad \text{(ii)}$$

VII.3 The Graph Γ.

3.1 Definition. The nodes of Γ are v.a.c. n-simplices of G lying in $\{x \in R^n | x_1 \geq 0\}$ and v.a.c. (n-1)-simplices of G^+ lying in $\{x \in R^n | x_1 = 0\}$. Two nodes of Γ are adjacent if one is a face of the other or if they share a v.a.c. facet.

Let us first examine the degrees of nodes of Γ. If τ is a v.a.c. (n-1)-simplex and a node of Γ, τ is a facet of just one simplex of G in $\{x \in R^n | x_1 \geq 0\}$; and clearly this simplex is v.a.c. Since no facet of τ can be v.a.c., τ has degree one. If σ is an n-simplex of G and a node of Γ, we distinguish the cases where σ is v.c. and where it is not.

If σ is v.c., the system (ii) of 2.3 has two basic feasible solutions, one of which is the solution of 2.3(i). Thus σ has just one v.a.c. facet τ. Whether τ lies in $\{x \in R^n | x_1 = 0\}$ (and hence is a node of Γ) or not (in which case, τ is a facet of another node σ' of Γ), we find that σ has degree 1.

If σ is not v.c., then the two basic feasible solutions to 2.3(ii) show that σ has exactly two v.a.c. facets. Hence in this case σ has degree 2.

We now see that Γ shares many desirable features with Γ_n of Chapter IV. To obtain an algorithm, we would like to know that Γ has only a finite number of nodes and has a natural starting node.

Let τ_0 be the (n-1)-simplex of G^{n-1} containing x^0.

3.2 Lemma.

(a) Γ has only a finite number of nodes.

(b) τ_0 is the only (n-1)-simplex that is a node of Γ.

Proof.

(a). Let σ be a node of Γ. If σ has a vertex in C, then $\sigma \subseteq B(C,\delta\sqrt{n})$, which is compact. Thus there are only a finite number of such simplices. Suppose that no vertex of σ lies in C. Then $L_\sigma = \begin{bmatrix} 1 & \ldots & 1 \\ c-y^0 & \ldots & c-y^k \end{bmatrix}$ where $\sigma = \langle y^0,\ldots,y^k \rangle$. Since σ is v.a.c., it is almost complete and there exist $w_0,\ldots,w_k \geq 0$ and $\lambda \geq 0$ with $\sum_{i=0}^k w_i + \lambda = 1$ and $\sum_{i=0}^k w_i(c-y^i) + \lambda d = 0$. We know that $\rho = \sum_{i=0}^k w_i > 0$, for otherwise $\lambda = 1$ and $d = 0$, a contradiction. Let $\mu_i = w_i/\rho$ and $\nu = \lambda/\rho \geq 0$. Then

$$\sum_{i=0}^k \mu_i y^i = c + \nu d,$$

i.e., $c + \nu d \in \bar{\sigma}$. Because $\bar{\sigma} \subseteq \{x \in R^n | x_1 \geq 0\}$, $\nu \leq 1$; thus $\bar{\sigma}$ meets $C' = \{c + \nu d | 0 \leq \nu \leq 1\}$ and $\sigma \subseteq B(C',\delta\sqrt{n})$. Since the latter set is compact, there is only a finite number of nodes of Γ with no vertices in C. The conclusion follows.

(b) Any (n-1)-simplex τ that is a node of Γ lies in $\{x \in R^n | x_1 = 0\}$ and has no vertex in C. As in (a), $\bar{\tau}$ must contain a point $c + \nu d$, $0 \leq \nu \leq 1$; and since its first coordinate must be zero, $\bar{\tau}$ must contain $c + d = x^0$. Hence $\tau = \tau_0$.

We must now show that τ_0 is v.a.c. Let $x^0 = 2\sum_{i\in N} w_i y^i$, where $\tau_0 = \langle y^1,\ldots,y^n \rangle$. Then $w > 0$ and $u^T w = \frac{1}{2}$. We show that $(w^T, \frac{1}{2})^T$ solves (*) of 2.1. Indeed, $w > 0$ and $\frac{1}{2} > 0$ with $u^T w + \frac{1}{2} = 1$, while

$$\sum_{i\in N} w_i \ell(y^i) + \frac{1}{2} d = \sum_{i\in N} w_i(c-y^i) + \frac{1}{2} d = \frac{1}{2} c - \frac{1}{2} x^0 + \frac{1}{2} d = 0.$$

Hence τ_0 is almost complete. Also, w and $\frac{1}{2}$ suffice as the first columns of W and e in 2.3(ii) and are strictly positive. If $L = [L_{\tau_0} (\begin{smallmatrix}1\\d\end{smallmatrix})]$ is nonsingular, we can obtain a solution to 2.3(ii). But certainly $y^1,\ldots,y^n$, $c-d$ are affinely independent; hence so are $d, c-y^1,\ldots,c-y^n$, proving L is nonsingular. □

Combining our analysis of the degrees of nodes of Γ with 3.2, we obtain

<u>3.3 Theorem</u>. Each connected component of the finite graph Γ is either a simple circuit or a simple path each of whose endpoints is either τ_0 or a v.c. simplex. □

VII.4 The Algorithm.

4.1. We trace the simple path in Γ with τ_0 as one endpoint.

Step 0. Let σ_1 be the unique n-simplex of G in $\{x \in R^n | x_1 \geq 0\}$ with τ_0 as a facet. Let y^+ be the vertex of σ_1 that is not a vertex of τ_0. Form the linear system (as in 3.2) showing τ_0 v.a.c. Set $m = 1$.

Step 1. Calculate the vector label $\ell(y^+)$ of y^+. Introduce the column $\binom{1}{\ell(y^+)}$ into the basis of the current lexicographic linear system. If the artificial column $\binom{1}{d}$ drops from the basis, STOP--σ_n is v.c. Otherwise, a unique column $\binom{1}{\ell(y^-)}$ drops, where y^- is a vertex of σ_m.

Step 2. Find the n-simplex σ_{m+1} of G containing all vertices of σ_m except y^-. Let y^+ be the new vertex. Set $m \leftarrow m + 1$ and return to Step 1.

4.2 Example (n = 2). Let $C = \{x \in R^2 | \frac{1}{2} \leq x_1, x_2 \leq 3\frac{1}{2}\}$ with $c = (1, 1\frac{1}{2})^T \in$ int C. Let $C_1 = \{x \in C | x_1, x_2 \leq 2\frac{1}{2}\}$, $C_2 = \{x \in C | x_1 \leq 2\frac{1}{2} \leq x_2\}$ and $C_3 = \{x \in C | x_1 \geq 2\frac{1}{2}\}$. Let $d^1 = (\frac{1}{2}, \frac{1}{4})^T$, $d^2 = (\frac{1}{2}, -\frac{1}{4})^T$ and $d^3 = (-\frac{1}{2}, 0)^T$. For $i = 1,2,3$, set $F_i(x) = \{x + d^i\}$ if $x \in C_i$, and $\emptyset$ otherwise; and set $F_4(x) = \{c\}$ if $x \in \partial C$, $\emptyset$ otherwise. Finally, let $F = \text{conv}(\bigcup_{i=1}^4 F_i)$. One can easily check that $(2\frac{1}{2}, 2\frac{1}{2})^T$ is the unique fixed point of F.

Pick $x^0 = (0, 1\frac{1}{2})^T$ so that $d = x^0 - c = (-1, 0)^T$. We will triangulate R^2 with J_1. Then $\tau^0 = \langle (0,1), (0,2) \rangle$. The relevant portion of R^2, together with the labels of the vertices shown by arrows, is indicated below.

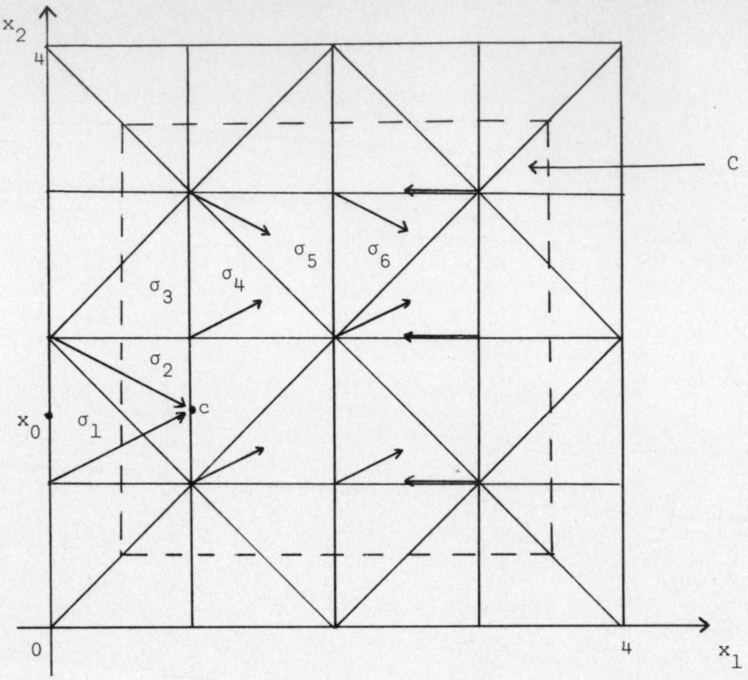

The initial simplex is $\sigma_1 = \langle (1,1)^T, (0,1)^T, (0,2)^T \rangle = \langle y^0, y^1, y^2 \rangle = j_1(y^0,\pi,s)$ with $\pi = (1,2)$ and $s = (-1,1)$. We have $\tau^0 = \langle y^1, y^2 \rangle$. Let $B = [L_{\tau^0}, (\frac{1}{d})] =$

$$\begin{bmatrix} 1 & 1 & 1 \\ 1 & 1 & -1 \\ \frac{1}{2} & -\frac{1}{2} & 0 \end{bmatrix}$$

be the current basis. Then

$$B^{-1} = \begin{bmatrix} \frac{1}{4} & \frac{1}{4} & 1 \\ \frac{1}{4} & \frac{1}{4} & -1 \\ \frac{1}{2} & -\frac{1}{2} & 0 \end{bmatrix} \begin{matrix} y^1 \\ y^2 \\ a \end{matrix}$$

; the first and second rows correspond to y^1 and y^2 and the third to the artificial column $(\frac{1}{d})$ (hence the notation "a"). Since the rows of B^{-1} are lexicographically positive, τ^0 is v.a.c. The iterations now proceed as follows.

Iteration m	Simplex σ_m	y^+	$f(y^+)$	$\ell(y^+)$	$B^{-1}\binom{1}{\ell(y^+)}$	y^-	New Basis Inverse	
1	$\langle y^0,y^1,y^2\rangle = \langle\binom{1}{1},\binom{0}{1},\binom{0}{2}\rangle$ $j_1(y^0,\pi,s)$, $\pi = (1,2)$, $s = (-1,1)$	$\begin{pmatrix}1\\1\end{pmatrix}$	$\begin{pmatrix}1\frac{1}{2}\\1\frac{1}{4}\end{pmatrix}$	$\begin{pmatrix}\frac{1}{2}\\\frac{1}{4}\end{pmatrix}$	$\begin{pmatrix}\frac{5}{8}\\\frac{1}{8}\\\frac{1}{4}\end{pmatrix}$	y^1	$\begin{bmatrix}\frac{2}{5} & \frac{2}{5} & \frac{8}{5}\\\frac{1}{5} & \frac{1}{5} & -\frac{6}{5}\\\frac{2}{5} & -\frac{3}{5} & -\frac{2}{5}\end{bmatrix}$	y^0 y^2 a
2	$\langle y^0,z^1,y^2\rangle = \langle\binom{1}{1},\binom{1}{2},\binom{0}{2}\rangle$ $j_1(y^0,\pi,s)$, $\pi = (2,1)$, $s = (-1,1)$	$\begin{pmatrix}1\\2\end{pmatrix}$	$\begin{pmatrix}1\frac{1}{2}\\2\frac{1}{4}\end{pmatrix}$	$\begin{pmatrix}\frac{1}{2}\\\frac{1}{4}\end{pmatrix}$	$\begin{pmatrix}1\\0\\0\end{pmatrix}$	y^0	["]	z^1 y^2 a
3	$\langle z^0,z^1,y^2\rangle = \langle\binom{1}{3},\binom{1}{2},\binom{0}{2}\rangle$ $j_1(z^0,\pi,s)$, $\pi = (2,1)$, $s = (-1,-1)$	$\begin{pmatrix}1\\3\end{pmatrix}$	$\begin{pmatrix}1\frac{1}{2}\\2\frac{3}{4}\end{pmatrix}$	$\begin{pmatrix}\frac{1}{2}\\-\frac{1}{4}\end{pmatrix}$	$\begin{pmatrix}\frac{1}{5}\\\frac{3}{5}\\\frac{1}{5}\end{pmatrix}$	y^2	$\begin{bmatrix}\frac{1}{3} & \frac{1}{3} & 2\\\frac{1}{3} & \frac{1}{3} & -2\\\frac{1}{3} & -\frac{2}{3} & 0\end{bmatrix}$	z^1 z^0 a
4	$\langle z^0,z^1,z^2\rangle = \langle\binom{1}{3},\binom{1}{2},\binom{2}{2}\rangle$ $j_1(z^0,\pi,s)$, $\pi = (2,1)$, $s = (1,-1)$	$\begin{pmatrix}2\\2\end{pmatrix}$	$\begin{pmatrix}2\frac{1}{2}\\2\frac{1}{4}\end{pmatrix}$	$\begin{pmatrix}\frac{1}{2}\\\frac{1}{4}\end{pmatrix}$	$\begin{pmatrix}1\\0\\0\end{pmatrix}$	z^1	["]	z^2 z^0 a
5	$\langle z^0,w^1,z^2\rangle = \langle\binom{1}{3},\binom{2}{3},\binom{2}{2}\rangle$ $j_1(z^0,\pi,s)$, $\pi = (1,2)$, $s = (1,-1)$	$\begin{pmatrix}2\\3\end{pmatrix}$	$\begin{pmatrix}2\frac{1}{2}\\2\frac{3}{4}\end{pmatrix}$	$\begin{pmatrix}\frac{1}{2}\\-\frac{1}{4}\end{pmatrix}$	$\begin{pmatrix}0\\1\\0\end{pmatrix}$	z^0	["]	z^2 w^1 a
6	$\langle w^0,w^1,z^2\rangle = \langle\binom{3}{3},\binom{2}{3},\binom{2}{2}\rangle$ $j_1(w^0,\pi,s)$, $\pi = (1,2)$, $s = (-1,-1)$	$\begin{pmatrix}3\\3\end{pmatrix}$	$\begin{pmatrix}2\frac{1}{2}\\3\end{pmatrix}$	$\begin{pmatrix}-\frac{1}{2}\\0\end{pmatrix}$	$\begin{pmatrix}\frac{1}{6}\\\frac{1}{6}\\\frac{2}{3}\end{pmatrix}$	-	$\begin{bmatrix}\frac{1}{4} & \frac{1}{2} & 2\\\frac{1}{4} & \frac{1}{2} & -2\\\frac{1}{2} & -1 & 0\end{bmatrix}$	z^2 w^1 w^0

Since the artificial column drops from the basis, σ_6 is very complete. From the first column of the basis inverse, the fixed point of f is $x^* = \frac{1}{4}z^2 + \frac{1}{2}w^1 + \frac{1}{2}w^0 = \begin{pmatrix}2\frac{1}{2}\\2\frac{3}{4}\end{pmatrix}$.

CHAPTER VIII: MERRILL'S ALGORITHM

VIII.1 An Extra Dimension. We will call the algorithms we have developed so far first-generation algorithms. They all suffer from computational inefficiency, caused by the following two features:

(i) we must choose a particular grid size with which to work throughout the algorithm; and

(ii) we must start outside the region of interest.

If the grid size is large, we obtain a poor approximation, while if it is small, we spend an inordinate amount of time moving through simplices .r from the solution.

One possible way to circumvent these difficulties is to refine the triangulation after finding a c.l. or v.c. simplex. The idea is illustrated with integer labels below:

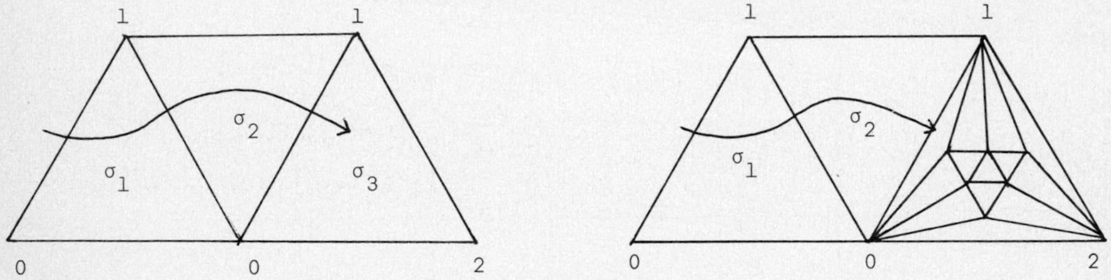

Unfortunately, the refined triangulation within σ_3 has simplices of as large diameter as σ_3 itself, and this characteristic cannot be eliminated without allowing escape from σ_3.

We will discuss two ways to avoid the problems of first-generation algorithms. The first method is that of Merrill and circumvents feature (ii) above; the algorithm can start anywhere. Feature (i) no longer causes problems, since we may apply the algorithm with a decreasing sequence of grid sizes, using the approximate solution just found as a starting point for the next application. Algorithms of this form we shall call second-generation algorithms; they include Merrill's restart method [48,49] and Kuhn and MacKinnon's sandwich method [41], discovered later but independently.

The second method, introduced by Eaves [14], eliminates features (i) and (ii) above. The grid size is automatically refined during the course of the algorithm. Such algorithms we will call third-generation, although they appeared at about the same time as Merrill's. These methods are most easily explained after one is familiar with Merrill's algorithm; they are the subject of the next chapter.

To motivate Merrill's algorithm, consider the following situation of Eaves' first algorithm:

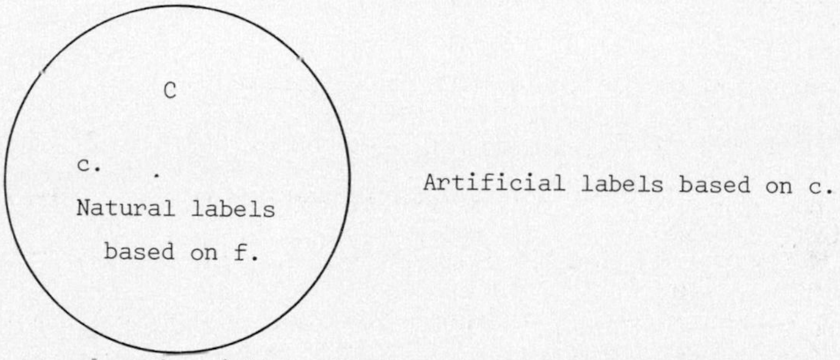

In order to solve the right problem, we had to label vertices in C naturally. To obtain a starting point, artificial labels were necessary and had to be outside C. Hence the starting point was outside C.

The artificial column d was needed for two reasons:

(i) to obtain the required degree of freedom in the linear system; and

(ii) to allow us to find a "deformed fixed point" (c+d) outside C and thus obtain a starting point.

If we want to start within C, we must have artificial labels within C, in conflict with our desire to have only natural labels in C. To solve the problem, we use two copies of C.

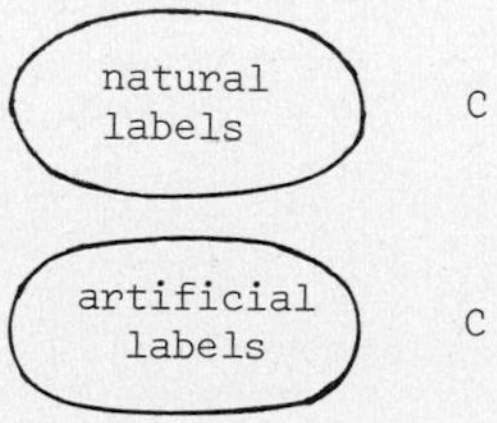

It is then natural to work in the (n+1)-dimensional set $[0,1] \times C$. Both reasons for d disappear. Since the dimension is increased by one, so is the number of vertices in a full-dimensional simplex, and we have the required degree of freedom. Also, we can start at a true fixed point of the artificial map of the bottom layer--it need not be deformed by translation by d.

It is clear why such algorithms are called sandwich methods--the two layers of the sandwich are $\{0\} \times C$ (artificial) and $\{1\} \times C$ (natural). (Eaves' first algorithm can be thought of as an "inefficient sandwich" where a hole is cut in a slice of bread for insertion of a piece of ham.) Using the same analogy, we can liken the third-generation algorithms to "club sandwiches":

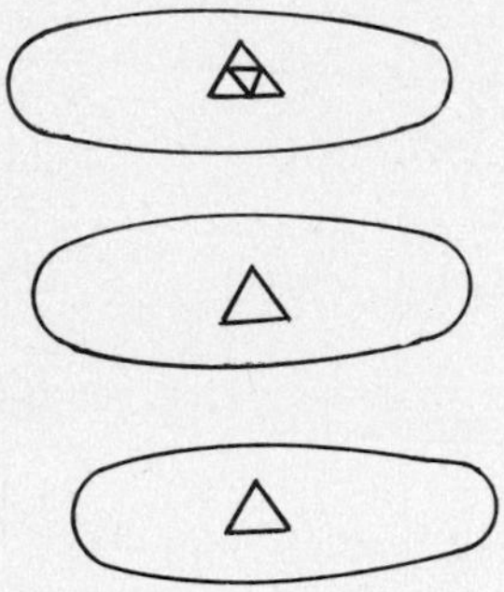

VIII.2 Preliminaries. We wish to find an approximate fixed point of a u.s.c. mapping $F_1 \colon R^n \to R^{n*}$. Assume that F_1 satisfies the hypotheses of V.3.4:

2.1 Assumption. There exist $x^0 \in R^n$, $\delta > 0$, and $\mu > 0$ such that whenever $x \notin B(x^0,\mu)$, $f \in F_1(x)$, and $z \in B(x,\delta)$, $(f-x)^T(x^0-z) > 0$.

Pick $c \in R^n$ arbitrarily, and let A be a non-singular $n \times n$ real matrix. Define $F_0 \colon R^n \to R^{n*}$ by $F_0(x) = \{x + A(c-x)\}$. For any $x = (x_0,x_1,\ldots,x_n)^T \in [0,1] \times R^n$, let $p(x) = (x_1,\ldots,x_n)^T$ be its projection in R^n. Then we can define $F \colon [0,1] \times R^n \to R^{n*}$ by $F(x) = x_0F_1(p(x)) + (1-x_0)F_0(p(x))$. We call x a fixed point of F if $F(x) = p(x)$. Note that $(0,c^T)^T$ is the only fixed point of F in $\{0\} \times R^n$ and that any fixed point of F in $\{1\} \times R^n$ projects into a fixed point of F_1.

Let G be a triangulation of $[0,1] \times R^n$ with $G^0 \subseteq \{0,1\} \times R^n$. Such a triangulation is called special. If $\sigma \in G$, $\mathrm{diam}'_q\sigma = \sup\{\|p(x) - p(z)\|_q \mid x,z \in \sigma\}$ is the diameter of the projection of σ, and $\mathrm{mesh}'_q G = \sup_{\sigma\in G}\mathrm{diam}'_q\sigma$ for $q = 2,\infty$.

The triangulation of R^n whose simplices are $\langle p(y^0),\ldots,p(y^n)\rangle$ for

$\sigma = \langle y^0,\dots,y^n\rangle \in G^n$ with $\sigma \subseteq \{i\} \times R^n$ is denoted G_i, $i = 0,1$.

2.2 Examples. The restrictions $\tilde{K}_1$ and $\tilde{J}_1$ of K_1 and J_1 (as triangulations of R^{n+1}) to $[0,1] \times R^n$ are special triangulations. Let E denote the $(n+1)\times(n+1)$ matrix $[v^0,\varepsilon v^1,\dots,\varepsilon v^n]$; then $\tilde{K}_1(\varepsilon) = \{\langle Ey^{-1},\dots,Ey^n\rangle | \langle y^{-1},\dots,y^n\rangle \in \tilde{K}_1\}$ and $\tilde{J}_1(\varepsilon)$, defined similarly, are special triangulations. $(\tilde{K}_1(\varepsilon))_0 = (\tilde{K}_1(\varepsilon))_1 = \varepsilon K_1$ and similarly for J_1. We have $\text{mesh}'_\infty(\tilde{K}_1(\varepsilon)) = \text{mesh}'_\infty(\tilde{J}_1(\varepsilon)) = \varepsilon$ and $\text{mesh}'_2(\tilde{K}_1(\varepsilon)) = \text{mesh}'_2(\tilde{J}_1(\varepsilon)) = \varepsilon\sqrt{n}$.

We will assume that c lies in an n-simplex of G_0 or, equivalently, that $(0,c^T)^T$ lies in an n-simplex τ_0 of G. (If not, c can be perturbed.)

2.3 Definition. Let ℓ be a vector labelling for F with respect to G. If $\sigma = \langle y^{-1},\dots,y^k\rangle \in G^{k+1}$, define the label matrix of σ to be

$$L_\sigma = \begin{bmatrix} 1 & \dots & 1 \\ \ell(y^{-1}) & \dots & \ell(y^k) \end{bmatrix}$$

We say σ is complete if there is a solution to

$$L_\sigma w = v^0, \quad w \geq 0, \qquad (*)$$

and very complete (v.c.) if there is a solution to

$$L_\sigma W = I, \quad W \geq 0. \qquad (**)$$

2.4 Lemma. Let F, G, and ℓ be as above and let $\sigma \in G^n \cup G^{n+1}$. Then there is a zero of ℓ in $\bar{\sigma}$ iff σ is complete. The only zero of ℓ in $\{0\} \times R^n$ is $(0,c^T)^T$, and the projection of any zero of ℓ in $\{1\} \times R^n$ is a fixed point of a p.l. approximation to F_1 with respect to G_1.

Proof. Similar to VII.1.1. □

Note that $\ell(y)$ depends only on c and A if $y \in \{0\} \times R^n$ and only on F_1 if $y \in \{1\} \times R^n$. For this reason many authors define two labelling rules. Our approach defines ℓ throughout $[0,1]\times R^n$; thus Merrill's algorithm can be seen to trace zeroes of ℓ from the artificial level to the natural level.

VIII.3 The Graph Γ.

3.1 Definition. The nodes of Γ are v.c. (n+1)-simplices of G and n-simplices of G^n lying in $\{0\}\times R^n$ or $\{1\}\times R^n$. Two nodes are adjacent if one is a face of the other or if they share a v.c. facet.

The following result shows that Γ has the right properties to be a basis for an algorithm.

3.2 Lemma.

(a) There is exactly one node of Γ lying in $\{0\}\times R^n$, namely, the n-simplex τ_0 of G^n containing $(0,c^T)^T$.

(b) Any node of Γ lying in $\{1\}\times R^n$ gives an approximate fixed point of F_1.

(c) If $\text{mesh}_2' G \leq \delta$ and A is positive definite, and assumption 2.1 holds, Γ has only finitely many nodes.

Proof. Parts (a) and (b) follow from 2.4; the fact that the simplex of G^n containing $(0,c^T)^T$ is v.c. follows as in VII.3.2.

(c) We show that outside a compact region there are no v.c. simplices. We have x^0, μ, and δ so that whenever $x \notin B(x^0,\mu)$, $f \in F_1(x)$, and $z \in B(x,\delta)$, $(f-x)^T(x^0-z) > 0$.

Let $\|A\|_2 = \max\{\|Ax\|_2 \mid \|x\|_2 = 1\}$ and $\eta = \min\{x^TAx \mid \|x\|_2 = 1\}$. Since $\{x \in R^n \mid \|x\|_2 = 1\}$ is compact, the optima are attained and $\|A\|_2 < \infty$, $\eta > 0$.

Let $\mu' = \max\{\mu + \delta, \|A\|_2(\delta + \|c-x^0\|_2)/\eta\}$. Let $C = [0,1]\times B(x^0,\mu') \subseteq R^{n+1}$. Then C is compact--we show that all complete simplices lie in C. Let $\sigma \in G$ have some vertex, y^*, say, outside C; let $z^* = p(y^*)$ and $s = x^0 - z^*$. We show that $s^T\ell(y) > 0$ for all vertices y of σ; hence σ cannot be complete.

Let y be a vertex of σ with $y_0 = 1$. Let $z = p(y)$; we have $\ell(y) = f - z$ for some $f \in F_1(z)$. Since $z^* \notin B(x^0,\mu+\delta)$ and $z \in B(z^*,\delta)$, $z \notin B(x^0,\mu)$. Hence $s^T\ell(y) = (f-z)^T(x^0-z^*) > 0$ by assumption 2.1.

Let y be a vertex of σ with $y_0 = 0$. Let $z = p(y)$--then $\ell(y) = A(c-z)$

and $s^T\ell(y) = (x^0-z^*)^T A(c-z) = (x^0-z^*)^T A(x^0-z^*) + (x^0-z^*)^T A(c-x^0) + (x^0-z^*)^T A(z^*-z)$.
The first term above is positive; we claim that it dominates the other two terms. Indeed, if $\|x^0-z^*\|_2 = \nu > \mu'$,

$$(x^0-z^*)^T A(x^0-z^*) \geq \nu^2\eta > \eta\nu\mu' \geq \eta\nu\|A\|_2(\delta + \|c-x^0\|_2)/\eta.$$

$\nu\|A\|_2\delta$ is at least $\nu\|A(z^*-z)\|_2$ and $\nu\|A\|_2\|c-x^0\|_2$ is at least $\nu\|A(c-x^0)\|_2$. This proves our claim; hence $s^T\ell(y) > 0$ in this case also.

Thus all complete simplices lie in C and the result is proved. □

Exactly as in VII.3 we obtain that nodes of Γ have degree 1 or 2 according to whether they are n-simplices or (n+1)-simplices.

3.3 Theorem. Each connected component of the finite graph Γ is either a simple circuit or a simple path each of whose endpoints is τ_0 or a v.c. n-simplex in $\{1\} \times R^n$. □

VIII.4 The Algorithm.

4.1 Merrill's Basic Algorithm. Suppose we have F_1, δ, G, and ℓ as in Section 2 with $\text{mesh}_2' G \leq \delta$. Let τ_0 be the simplex of G^n containing $(0,c^T)^T$ and σ_1 the simplex of G^{n+1} with τ_0 as a facet.

Step 0. Form the lexicographic linear system showing τ_0 v.c. Let y^+ be the vertex of σ_1 not in τ_0. Set $m \leftarrow 1$.

Step 1. Calculate $\ell(y^+)$ and introduce the column $\binom{1}{\ell(y^+)}$ into the current basis, displacing a unique column $\binom{1}{\ell(y^-)}$ with y^- a vertex of σ_m.

Step 2. If the facet τ of σ_m opposite y^- lies in $\{1\} \times R^n$, STOP--$\bar{\tau}$ contains an approximate fixed point of F_1. Otherwise, let σ_{m+1} be the unique simplex of G sharing the facet τ with σ_m. Let y^+ be the new vertex of σ_{m+1}, set $m \leftarrow m+1$, and return to Step 1.

4.2 The Restart Algorithm.

Step 0. Pick a sequence $\delta_k \downarrow 0$ with $\delta_0 \leq \delta$. For each k pick a special triangulation $G_{(k)}$ with $\text{mesh}' G_{(k)} = \delta_k$. Choose $x^0 \in R^n$ and set $p \leftarrow 0$.

Step 1. Apply algorithm 4.1 with $G_{(p)}$ replacing G, x^p replacing c and some positive definite matrix A_p to obtain an approximate fixed point x^{p+1}. Set $p \leftarrow p+1$ and return to Step 1.

Of course, x^p must be perturbed if it does not lie in an n-simplex of $(G_{(p)})_0$.

One typically takes $G_{(k)} = \tilde{K}_1(\alpha^k \delta_0)$ where $0{\cdot}1 \leq \alpha \leq 0{\cdot}6$. We choose α small for "smooth" functions and large for highly nonlinear functions or mappings.

VIII.5 An Example. Let $n = 2$ and $F_1(x) = \left\{ x + \begin{pmatrix} 10 & 0 \\ 0 & 0 \end{pmatrix} ((\frac{2}{3},\frac{1}{3})^T - x) \right\}$. Clearly, F_1 has just one fixed point at $(\frac{2}{3},\frac{1}{3})^T$. Let us choose $c = (\frac{2}{3},\frac{1}{3})^T$ (a perfect guess) and $A = \begin{pmatrix} 1 & 0 \\ 0 & 1 \end{pmatrix}$. We will apply algorithm 4.1 using $\tilde{K}_1$.

We find $\tau^0 = \langle y^{-1}, y^0, y^1 \rangle = \left\langle \begin{pmatrix} 0 \\ 0 \\ 0 \end{pmatrix}, \begin{pmatrix} 0 \\ 1 \\ 0 \end{pmatrix}, \begin{pmatrix} 0 \\ 1 \\ 1 \end{pmatrix} \right\rangle$. The corresponding label matrix is

$$L_{\tau^0} = \begin{bmatrix} 1 & 1 & 1 \\ \frac{2}{3} & -\frac{1}{3} & -\frac{1}{3} \\ \frac{1}{3} & \frac{1}{3} & -\frac{2}{3} \end{bmatrix}; \text{ this is the initial basis, with inverse } \begin{bmatrix} \frac{1}{3} & 1 & 0 \\ \frac{1}{3} & -1 & 1 \\ \frac{1}{3} & 0 & -1 \end{bmatrix} \begin{matrix} y^{-1} \\ y^0 \\ y^1 \end{matrix}.$$

Since the rows are lexicographically positive, τ^0 is v.c. The iterations now proceed as follows.

Iteration m	Simplex σ_m	y^+	$\ell(y^+)$	$B^{-1}\begin{pmatrix}1\\ \ell(y^+)\end{pmatrix}$	y^-	New Basis Inverse	
1	$\langle y^{-1},y^0,y^1,y^2\rangle = \left\langle \begin{pmatrix}0\\0\\0\end{pmatrix},\begin{pmatrix}0\\1\\0\end{pmatrix},\begin{pmatrix}0\\1\\1\end{pmatrix},\begin{pmatrix}1\\1\\1\end{pmatrix}\right\rangle$ $k_1(y^{-1},\pi), \quad \pi = (1,2,0)$	$\begin{pmatrix}1\\1\\1\end{pmatrix}$	$\begin{pmatrix}-\frac{10}{3}\\ -\frac{2}{3}\end{pmatrix}$	$\begin{pmatrix}-3\\3\\1\end{pmatrix}$	y^0	$\begin{bmatrix}\frac{2}{3} & 0 & 1\\ \frac{1}{9} & -\frac{1}{3} & \frac{1}{3}\\ \frac{2}{9} & \frac{1}{3} & -\frac{4}{3}\end{bmatrix}$	y^{-1} y^2 y^1
2	$\langle y^{-1},z^0,y^1,y^2\rangle = \left\langle \begin{pmatrix}0\\0\\0\end{pmatrix},\begin{pmatrix}0\\0\\1\end{pmatrix},\begin{pmatrix}0\\1\\1\end{pmatrix},\begin{pmatrix}1\\1\\1\end{pmatrix}\right\rangle$ $k_1(y^{-1},\pi), \quad \pi = (2,1,0)$	$\begin{pmatrix}0\\0\\1\end{pmatrix}$	$\begin{pmatrix}\frac{2}{3}\\ -\frac{2}{3}\end{pmatrix}$	$\begin{pmatrix}0\\ -\frac{1}{3}\\ \frac{4}{3}\end{pmatrix}$	y^1	$\begin{bmatrix}\frac{2}{3} & 0 & 1\\ \frac{1}{6} & -\frac{1}{4} & 0\\ \frac{1}{6} & \frac{1}{4} & -1\end{bmatrix}$	y^{-1} y^2 z^0
3	$\langle y^{-1},z^0,z^1,y^2\rangle = \left\langle \begin{pmatrix}0\\0\\0\end{pmatrix},\begin{pmatrix}0\\0\\1\end{pmatrix},\begin{pmatrix}1\\0\\1\end{pmatrix},\begin{pmatrix}1\\1\\1\end{pmatrix}\right\rangle$ $k_1(y^{-1},\pi), \quad \pi = (2,0,1)$	$\begin{pmatrix}1\\0\\1\end{pmatrix}$	$\begin{pmatrix}\frac{20}{3}\\ -\frac{2}{3}\end{pmatrix}$	$\begin{pmatrix}0\\ -\frac{3}{2}\\ \frac{5}{2}\end{pmatrix}$	z^0	$\begin{bmatrix}\frac{2}{3} & 0 & 1\\ \frac{4}{15} & -\frac{1}{10} & -\frac{3}{5}\\ \frac{1}{15} & \frac{1}{10} & -\frac{2}{5}\end{bmatrix}$	y^{-1} y^2 z^1
4	$\langle y^{-1},w^0,z^1,y^2\rangle = \left\langle \begin{pmatrix}0\\0\\0\end{pmatrix},\begin{pmatrix}1\\0\\0\end{pmatrix},\begin{pmatrix}1\\0\\1\end{pmatrix},\begin{pmatrix}1\\1\\1\end{pmatrix}\right\rangle$ $k_1(y^{-1},\pi), \quad \pi = (0,2,1)$	$\begin{pmatrix}1\\0\\0\end{pmatrix}$	$\begin{pmatrix}\frac{20}{3}\\ \frac{1}{3}\end{pmatrix}$	$\begin{pmatrix}1\\ -\frac{3}{5}\\ \frac{3}{5}\end{pmatrix}$	z^1	$\begin{bmatrix}\frac{5}{9} & -\frac{1}{6} & \frac{5}{3}\\ \frac{1}{3} & 0 & -1\\ \frac{1}{9} & \frac{1}{6} & -\frac{2}{3}\end{bmatrix}$	y^{-1} y^2 w^0
5	$\langle y^{-1},w^0,w^1,y^2\rangle = \left\langle \begin{pmatrix}0\\0\\0\end{pmatrix},\begin{pmatrix}1\\0\\0\end{pmatrix},\begin{pmatrix}1\\1\\0\end{pmatrix},\begin{pmatrix}1\\1\\1\end{pmatrix}\right\rangle$ $k_1(y^{-1},\pi), \quad \pi = (0,1,2)$	$\begin{pmatrix}1\\1\\0\end{pmatrix}$	$\begin{pmatrix}-\frac{10}{3}\\ \frac{\text{[illegible]}}{3}\end{pmatrix}$	$\begin{pmatrix}\frac{5}{3}\\ 0\\ -\frac{2}{3}\end{pmatrix}$	y^{-1}	$\begin{bmatrix}\frac{1}{3} & -\frac{1}{10} & 1\\ \frac{1}{3} & 0 & -1\\ \frac{1}{3} & \frac{1}{10} & 0\end{bmatrix}$	w^1 y^2 w^0

The facet τ of σ_5 opposite y^{-1} lies in $\{1\} \times R^2$; thus τ is v.c. and gives as an approximate fixed point of F_1 the point $x^* = \frac{1}{3}\,p(w^1) + \frac{1}{3}\,p(y^2) + \frac{1}{3}\,p(w^0) = \begin{pmatrix}\frac{2}{3}\\ \frac{1}{3}\end{pmatrix}$. Since F_1 is a linear function, x^* is a fixed point of F_1.

The course of the algorithm is shown in the following diagram. The dotted line is the locus of zeroes of ℓ.

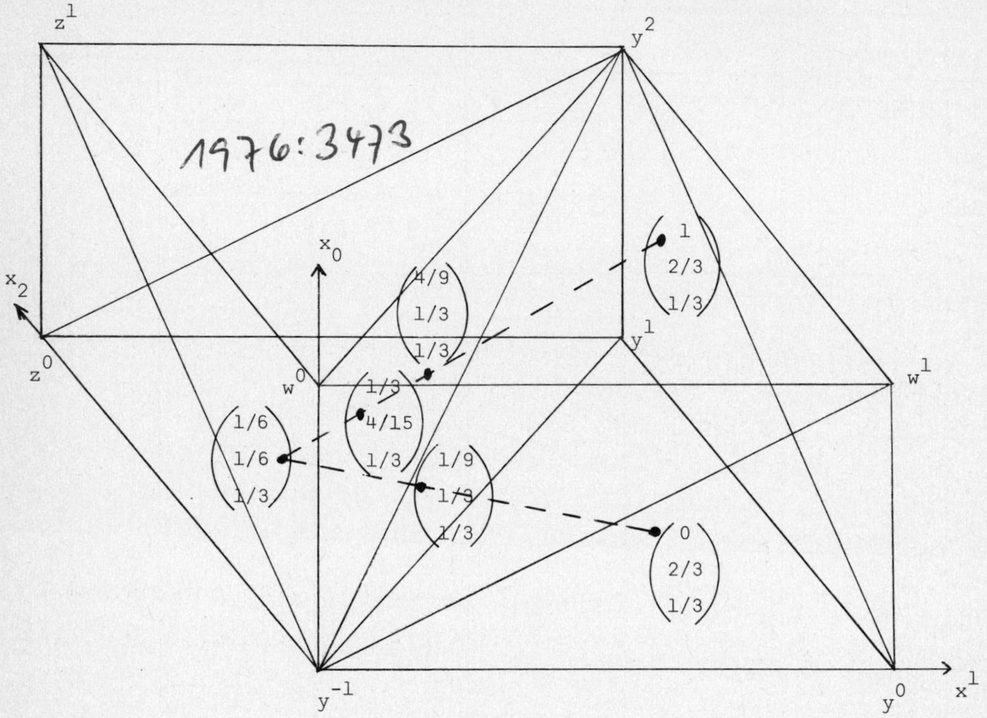

This problem required 4 evaluations of a point of $F_1(x)$ and one evaluation of $A(c-x)$, for a total of 5 linear programming pivot steps. In contrast, if we had taken $A = \begin{pmatrix} 10 & 0 \\ 0 & 1 \end{pmatrix}$ (so that the artificial map F_0 coincides with F_1), only 3 function evaluations and 3 linear programming pivots would have been required. Indeed, the simplices encountered are now $\sigma_1' = \langle y^{-1}, y^0, y^1, y^2 \rangle$, $\sigma_2' = \langle y^{-1}, y^0, w^1, y^2 \rangle$, and $\sigma_3' = \langle y^{-1}, w^0, w^1, y^2 \rangle$; to see this, note that now $\ell\binom{1}{x} = \ell\binom{0}{x}$ for any x.

Clearly, to obtain a v.c. simplex in $\{1\} \times R^2$, at least three pivots are necessary (in general, we need $n+1$). The example illustrates this important point: to obtain fast convergence, it is not enough to make a good guess for c--we must also make a good guess for A. A should be close to the Jacobian of g if g is differentiable and $F_1(x) = \{x + g(x)\}$. In the early stages of applying algorithm 4.2 little information about this Jacobian is available, and A must be positive definite to assure convergence. (Usually, A is taken to be the identity.) Later, when fairly good approximate points have been generated, there is less concern that the algorithm will diverge--then we may take A to be an approximation to the Jacobian, even if it is not positive definite. The way to obtain such an approximation without much additional work is the subject of exercise 7.5. The use of the Jacobian was first proposed in [55]; other references on its use are [20], [27], [66], [67] and [74]. The method for obtaining an approximate Jacobian appeared in [66] and [53].

VIII.6 Integer Labelling. If $F_1 = \{f^1\}$ where f^1 is a continuous function, we can use Merrill's strategy together with integer labelling. We need a flexible artificial labelling to assure one and only one completely labelled simplex in $\{0\} \times R^n$. In fact, the following scheme achieves our goals. Define ℓ as before. Then set the integer label of y to be 0 if $\ell(y) < 0$, and otherwise to be $\min\{j \in N \mid \ell_j(y) = \max_k \ell_k(y)\}$.

To assure that the algorithm does not diverge, we must assume that every point outside a bounded region has a neighborhood of radius δ in which some integer label is excluded. Since we want this condition to hold for any starting point c, we require the following.

6.1 Assumption. There exist $x^0 \in R^n$ and $\delta > 0$ such that corresponding to every $\nu > 0$ there is some $\mu > 0$ with every $x \notin B(x^0,\mu)$ satisfying:

either (a) $x \not\geq x^0 - \nu u$ and $f^1(z) \not\leq z$ for all $z \in B(x,\delta)$; or

(b) for some $i \in N$, $x_i > x_i^0 + \nu$ and $f_i^1(z) - z_i < \max_k(f_k^1(z) - z_k)$ for all $z \in B(x,\delta)$.

Under this assumption algorithm 4.1, using integer labelling and a special triangulation G with mesh $G \leq \delta$, cannot diverge. For if μ corresponds to $\nu = \delta + \|x^0-c\|_\infty$, any simplex outside $[0,1] \times B(x^0,\mu)$ is missing label 0 (in case (a)) or label i (in case (b)). We omit the details. See [20] and [27] for related labelling rules.

VIII.7 Exercises.

7.1. Let $L_\sigma = \begin{bmatrix} 1 & \cdots & 1 \\ c-y^0 & \cdots & c-y^n \end{bmatrix}$ where $\sigma = \langle y^0,\ldots,y^n\rangle$ is an n-simplex. Find L_σ^{-1} explicitly for $\sigma = k_1(y^0,\pi)$ and $\sigma = j_1(y^0,\pi,s)$.

7.2. Apply algorithm 4.1 exactly as in Section 5 but with $F_1(x) = \{x + \begin{pmatrix} 4 & 0 \\ 0 & 1 \end{pmatrix}(c-x)\}$. Note: degeneracy occurs and lexicographic rules must be used.

7.3. Apply the integer labelling version of algorithm 4.1 to the example of Section 5 and with $F_1(x) = \{x + \begin{pmatrix} 1 & 0 \\ 0 & 10 \end{pmatrix}(c-x)\}$. Note that in the latter case the final simplex does not contain the fixed point of F_1.

7.4. State and prove a result giving the degree of approximation obtained when one uses the integer labelling of Section 6.

7.5. Imagine you are trying to find a fixed point of f, where $f(x) = \begin{pmatrix} 2 & 1 & -2 \\ 2 & 3 & -3 \\ 4 & 3 & -5 \end{pmatrix} x + \begin{pmatrix} 2 \\ 3 \\ 3 \end{pmatrix}$, which is equivalent to finding a zero of g with $g(x) = f(x) - x = Ax + b$, $A = \begin{pmatrix} 1 & 1 & -2 \\ 2 & 2 & -3 \\ 4 & 3 & -6 \end{pmatrix}$, $b = \begin{pmatrix} 2 \\ 3 \\ 3 \end{pmatrix}$. Somehow you have found the simplex in $2K_1$ given by $\sigma = \langle (2,-4,0)^T,(2,-2,0)^T,(2,-2,2)^T,(4,-2,2)^T \rangle$ $k_1((2,-4,0)^T, (2,3,1))$. The label matrix associated with σ is

$$B = \begin{bmatrix} 1 & 1 & 1 & 1 \\ 0 & 2 & -2 & 0 \\ -1 & 3 & -3 & 1 \\ -1 & 5 & -7 & 1 \end{bmatrix} \qquad \text{with } B^{-1} = \begin{bmatrix} \frac{1}{2} & 0 & -1 & \frac{1}{2} \\ 0 & 1 & \frac{1}{2} & -\frac{1}{2} \\ 0 & \frac{1}{2} & \frac{1}{2} & -\frac{1}{2} \\ \frac{1}{2} & -\frac{3}{2} & 0 & \frac{1}{2} \end{bmatrix}.$$

Since the first column of B^{-1} $\left(= B^{-1}\begin{pmatrix} 1 \\ 0 \\ 0 \\ 0 \end{pmatrix}\right)$ is nonnegative, we have a linear approximate fixed point of f, given by $\frac{1}{2}y^0 + \frac{1}{2}y^3 = (3,-3,1)^T$. Since f is linear, this is in fact a fixed point of f. However, much more information can be extracted from B^{-1} (which is available to us if a fixed-point algorithm gave σ).

First discard the first column of B^{-1}, then premultiply the rest by

$R = \begin{bmatrix} 0 & 1 & 1 & 1 \\ 0 & 0 & 1 & 1 \\ 0 & 0 & 0 & 1 \end{bmatrix}$ to get $C = \begin{bmatrix} 0 & 1 & -\frac{1}{2} \\ -1 & \frac{1}{2} & 0 \\ -\frac{3}{2} & 0 & \frac{1}{2} \end{bmatrix}$. Now put the jth row of C into the $\pi(j)$'th position to get $D = \begin{bmatrix} -\frac{3}{2} & 0 & \frac{1}{2} \\ 0 & 1 & -\frac{1}{2} \\ -1 & \frac{1}{2} & 0 \end{bmatrix}$. Finally, multiply all entries in D by the grid size $\delta = 2$ to get $E = \begin{bmatrix} -3 & 0 & 1 \\ 0 & 2 & -1 \\ -2 & 1 & 0 \end{bmatrix}$. Now note that $E = A^{-1}$!

Find out what operations on B give in turn C^{-1}, D^{-1}, and finally E^{-1}. Now apply these operations when B has the general form: $\begin{bmatrix} 1 & \cdots & 1 \\ g(y^0) & \cdots & g(y^n) \end{bmatrix}$ and $\langle y^0,\ldots,y^n \rangle = k_1(y^0,\pi)$, a simplex of δK_1. Conclude that the inverse of the label matrix can be used to find the inverse of an approximation to the Jacobian of g.

CHAPTER IX: HOMOTOPY ALGORITHMS

The third-generation algorithms we consider here are based on the work of Eaves [14] and Eaves and Saigal [16]. To motivate the methods, consider a continuous function $f: S^1 \to S^1$. Instead of triangulating S^1, we triangulate S^2 and omit one simplex--S^2_2 is identified with S^1:

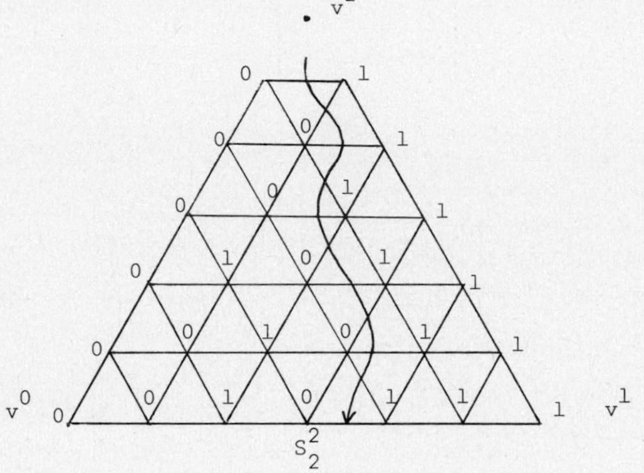

Label each vertex 0 or 1 according to whether its projection from v^2 onto $S^2_2 \equiv S^1$ has label 0 or 1. Then trace the natural path from the top 1-simplex down to S^1. The horizontal levels can be considered as successively finer triangulations of S^1. These triangulations have meshes $1, \frac{1}{2}, \frac{1}{3}, \frac{1}{4}, \frac{1}{5}, \frac{1}{6}$ above.

As another example, an alternative way to view the well-known bisection algorithm is by paths in triangulations:

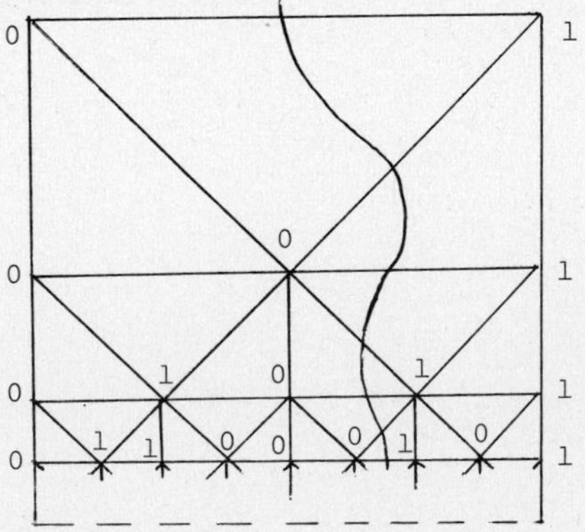

Here the successive triangulations of S^1 have meshes $1, \frac{1}{2}, \frac{1}{4}, \frac{1}{8} \dots$.

In both cases above, we used integer labelling for simplicity. However, if vector labelling is used, then the path links fixed points of p.l. approximations to

f with successively finer grids. The triangulation is used to deform f from a linear function to a p.l. approximation to f.

IX.1 Homotopies and Triangulations with Continuous Refinement of Grid Size.

1.1 Definition. Let $f^i: R^n \to R^n$ be continuous functions for $i = 0,1$. Then $h: [0,1] \times R^n \to R^n$ is a homotopy from f^0 to f^1 if h is continuous and $h(i,x^T)^T = f^i(x)$ for all $x \in R^n$, $i = 0,1$.

Note that the vector label ℓ of Chapter VIII was a homotopy from $A(c-\cdot)$ to a piecewise linear approximation to F_1 minus the identity. We traced zeroes of this function ℓ. We would like our algorithms to trace fixed points of a homotopy from a given linear function to the function of interest. For computation these homotopies should be piecewise linear.

Let $f: R^n \to R^n$ and $F = \{f\}: R^n \to R^{n*}$. The constructions we will make are motivated by F of this form but are also useful if $\Gamma: R^n \to R^{n*}$ is merely u.s.c. We will note in brackets the differences for the latter case.

The basic idea is to consider different levels on which are defined better and better approximations to f:

level 0 _______________ f^0, a linear function

level 1 _______________ f^1, a p.l. approx. to F with respect to a coarse triangulation

level 2 _______________ f^2, a p.l. approx. to F with respect to a finer triangulation

level ∞ _______________ f itself

[Note: if $F \neq \{f\}$, level ∞ is omitted.]

It is convenient to put the levels in a geometric setting by considering level k as $R^n(k) = \{2^{-k}\} \times R^n$; denote by $R^n(\infty)$ the set $\{0\} \times R^n$.

We now define the triangulations we use to construct the homotopies.

1.2 Definition. Let $\delta_k \searrow 0$. Let G be a triangulation of $(0,1] \times R^n$ satisfying

(i) if $y \in G^0$, $y_0 = 2^{-k}$ for $k = 0,1,2,\ldots$;

(ii) for $k = 0,1,2,\ldots$ $G_k = \{\sigma \in G^n \mid \sigma \subseteq R^n(k)\}$ triangulates $R^n(k)$;

(iii) if $\sigma \subseteq (0,2^{-k}] \times R^n$, $\text{diam}'\sigma \leq \delta_k$ for $k = 0,1,2,\ldots$.

Then G is called a triangulation with continuous refinement of grid size (abbreviated to "a triangulation with crogs").

1.3 Examples ($n = 1$).

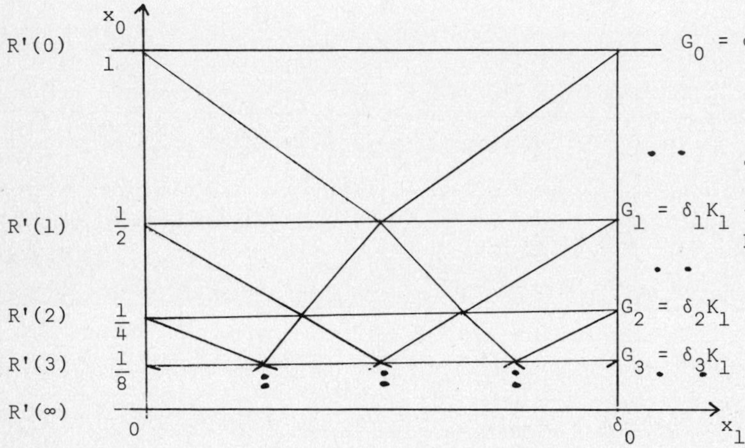

$\delta_K = \delta_0/(k+1)$

This triangulation is similar to Eaves' K_1 [14].

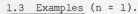

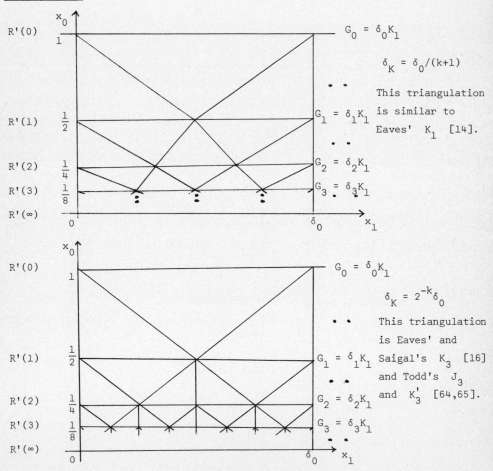

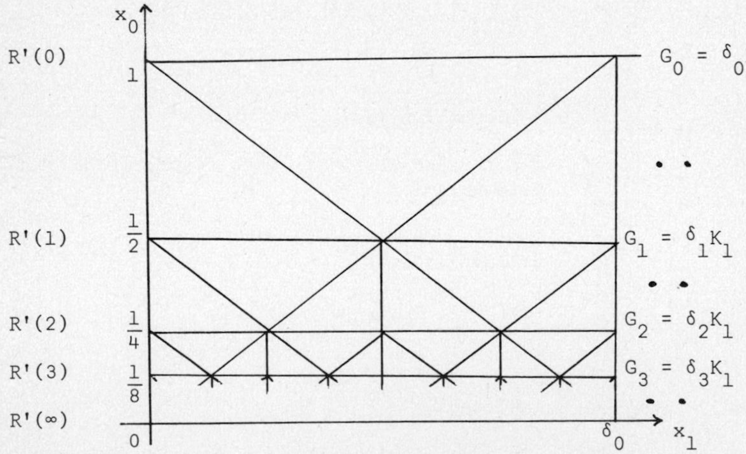

$\delta_K = 2^{-k}\delta_0$

This triangulation is Eaves' and Saigal's K_3 [16] and Todd's J_3 and K_3' [64,65].

1.4 Lemma. If G is a triangulation with crogs and $\sigma \in G$, then $\sigma \subseteq [2^{-k-1},2^{-k}] \times R^n$ for some $k \geq 0$.

Proof. Assume the contrary, i.e., that σ has vertices $y \in R^n(p)$ and $y' \in R^n(\ell)$, with $p < k < \ell$ three integers. Then some convex combination z of y and y' lies in $R^n(k)$. Thus $z \in \tau$ for some face τ of σ. This contradicts the fact that $z \in \rho$ for ρ some face of $\sigma' \in G_k$, from 1.2(ii). □

Using C we can construct a homotopy h from a linear function f^0 to f. Let $\alpha: R_+ \to R_+$ be a nondecreasing function, and let $f^0(x) = A(c-x) + x$.

<u>1.5 Definition</u>. Define $F': (0,1] \times R^n \to R^{n*}$ by $F'(x) = \{f^0(p(x))\}$ if $-\log_2 x_0 \leq \alpha(\|p(x) - c\|_2)$ and $F'(x) = F(p(x))$ otherwise. Restricted to $(0,1] \times R^n$, h is a piecewise linear approximation to F' with respect to G. Set $h(0,x^T)^T = f(x)$; thus $h: [0,1] \times R^n \to R^n$. [If $F \neq \{f\}$, omit the last sentence; we have $h: (0,1] \times R^n \to R^n$.]

We usually set α identically to zero; then we label according to f^0 on $R^n(0)$ and F on all other levels.

<u>1.6 Lemma</u>. h is a homotopy from f to f^0. [1.6 is meaningless if $F \neq \{f\}$.]

<u>Proof</u>. Clearly, $h(0,x^T)^T = f(x)$ and $h(1,x^T)^T = f^0(x)$ since f^0 is linear and agrees with its p.l. approximation. Also, h is continuous on $(0,1] \times R^n$ because it is a p.l. approximation. So let $x \in R^n(\infty)$ and $\varepsilon > 0$ be given. The continuity of f gives $\delta > 0$ such that $\|f(w) - f(p(x))\|_2 < \varepsilon$ whenever $w \in B(p(x),\delta)$. There exists $k > \alpha(\|p(x) - c\| + \delta)$ so that $\delta_k \leq \delta/2$. Let $M = [0,2^{-k-1}] \times B(p(x),\delta/2)$ and $z \in M$. We claim that $\|h(z) - h(x)\|_2 \leq \varepsilon$.

If $z_0 = 0$, $h(z) = f(p(z))$ and the claim is true. If $z_0 > 0$, z lies in $\bar{\sigma}$ for $\sigma = \langle y^{-1},\ldots,y^n \rangle \in G$. By 1.4, $y_0^i \leq 2^{-k}$ for all i and hence 1.2(iii) gives $\text{diam}'\sigma \leq \delta_k \leq \delta/2$. Thus $\|p(y^i) - p(x)\|_2 \leq \delta/2 + \delta/2 = \delta$ and $\|h(y^i) - h(x)\|_2 = \|f(p(y^i)) - f(p(x))\|_2 \leq \varepsilon$ for all i. The claim follows. □

<u>1.7 Definition</u>. We call $x \in [0,1] \times R^n$ a fixed point of h if $h(x) = p(x)$. Thus $x \in R^n(0)$ is a fixed point of h iff $p(x) = c$ or $x = (1,c^T)^T$ when A is nonsingular. Also, $x \in R^n(\infty)$ is a fixed point of h iff $p(x)$ is a fixed point of F. [The last statement is meaningless if $F \neq \{f\}$.]

A finite algorithm cannot obtain a fixed point of h in $R^n(\infty)$. The following result is therefore of interest.

1.8 Lemma. Let $\{x^k\}$ be a sequence of fixed points of h tending to $x \in R^n(\infty)$. Then $p(x)$ is a fixed point of F.

Proof. If $F = \{f\}$, we have $h: [0,1] \times R^n \to R^n$ continuous, so $h(x) = \lim h(x^k) = \lim p(x^k) = p(x)$. [Otherwise, the conclusion follows from 1.2(iii) and the method of proof of Kakutani's theorem.] □

IX.2 Complete Simplices.

2.1 Definition. Define $\ell: [0,1] \times R^n \to R^n$ by $\ell(x) = h(x) - p(x)$. [If $F \neq \{f\}$, $\ell: (0,1] \times R^n \to R^n$ is defined similarly.] If $\sigma \in G^n \cup G^{n+1}$, the label matrix of σ is

$$L_\sigma = \begin{bmatrix} 1 & \cdots & 1 \\ \ell(y^{-1}) & \cdots & \ell(y^k) \end{bmatrix} \quad \text{where} \quad \sigma = \langle y^{-1}, \ldots, y^k \rangle.$$

We call σ complete (or very complete) iff there is a feasible solution to $L_\sigma w = v^0$, $w \geq 0$ ($L_\sigma W = I$, $W \succeq 0$).

2.2 Lemma. Let $\sigma \in G^n \cup G^{n+1}$. Then there is a fixed point of h in $\bar{\sigma}$ iff σ is complete. If $\sigma \in G_0$ is complete, $(1,c^T)^T \in \bar{\sigma}$.

Proof. Trivial.

IX.3 The Graph Γ.

3.1 Definition. The nodes of Γ are v.c. (n+1)-simplices and v.c. n-simplices of G_0, with two such adjacent if one is a face of the other or if they share a v.c. facet.

Assume that $(1,c^T)^T$ lies in an n-simplex τ_0 of G_0. Then it follows that τ_0 is the unique node of Γ that is an n-simplex, and Γ contains an infinite simple path with τ_0 as its endpoint. The algorithm will follow this path. We want to assure that the path eventually penetrates every level.

Assume that F satisfies assumption 2.1 of Chapter VIII with x^0, μ and

$\delta > 0$. Suppose A is positive definite. Then as in VIII.3.2 no complete simplex of diameter at most δ can lie outside $C_{\mu'} = (0,1] \times B(x^0,\mu')$ for some μ'. Thus if G is a triangulation with crogs with meshes $\delta_0, \delta_1, \ldots$ and $\delta_0 \leq \delta$, all nodes of Γ lie in C. By appropriate use of the function α in 1.5 we obtain a stronger result.

3.2 Lemma. Under the conditions above, with $\delta_0 \leq \delta$, all nodes of Γ lie in $C_{\mu'}$. Furthermore, if α is an increasing function, then even with $\delta_0 > \delta$, all nodes of Γ lie in $C_{\mu''}$ for some $\mu'' > 0$.

Proof. If $\delta_0 \leq \delta$, the proof follows that of VIII.3.2. Otherwise, pick k so that $\delta_k \leq \delta$ and $\mu'' > \mu'$ so that $\alpha(\mu'' - \|x^0 - c\|_2 - \delta_0) \geq k+1$. Let σ be a simplex with some vertex outside $C_{\mu''}$. If $\sigma \subseteq (0,2^{-k}] \times R^n$, σ is not complete as in VIII.3.2. If $\sigma \subseteq [2^{-k},1] \times R^n$, $x \in \bar{\sigma} \Rightarrow -\log_2 x_0 < k+1$ and hence each vertex y of σ is labelled $A(c-p(y))$. Thus if σ is complete, it contains $(\lambda, c^T)^T$ for some λ, which contradicts $\sigma \not\subseteq C_{\mu''}$. □

Note that the same idea can be used in Merrill's algorithm to assure convergence even if $\delta_0 > \delta$.

IX.4 The Algorithm.

We trace the infinite path of Γ from τ_0 until a sufficiently accurate approximate fixed point is found. By 3.2 any infinite sequence of distinct, very-complete simplices must leave $[2^{-k},1] \times R^n$ for any k; by 2.2 we obtain a fixed point of h in $R^n(k)$ for every k; and, by 1.8, any cluster point of these fixed points is a fixed point of F.

Step 0. Let σ_1 be the unique (n+1)-simplex of G with τ_0 as a facet. Let y^+ be the vertex of σ_1 outside $R^n(0)$. Form the lexicographic linear system showing τ_0 very complete. Set $m \leftarrow 1$.

Step 1. Calculate $\ell(y^+)$. Introduce the column $\binom{1}{\ell(y^+)}$ into the current basis of the lexicographic linear system. Assume that the column $\binom{1}{\ell(y^-)}$ drops from the basis, with y^- a vertex of σ_m.

Step 2. Let τ_m be the very-complete facet of σ_m opposite y^-. If desired, calculate the fixed point of h in $\bar{\tau}_m$ and terminate if it is sufficiently accurate. Let σ_{m+1} be the simplex of G sharing the facet τ_m with σ_m. Let y^+ be the new vertex, set $m \leftarrow m+1$, and return to Step 1.

Generally, the fixed point of h in $\bar{\tau}_m$ is calculated only when $\tau_m \subseteq R^n(k)$ for some k.

IX.5 A Comparison with Merrill's Algorithm. The advantages of a homotopy algorithm over Merrill's algorithm are:

1) The function is continuously deformed rather than set to a constant function for each restart; and
2) The simplices generated can move from fine triangulations back to coarse ones, thus permitting a move to a more promising region in large simplices.

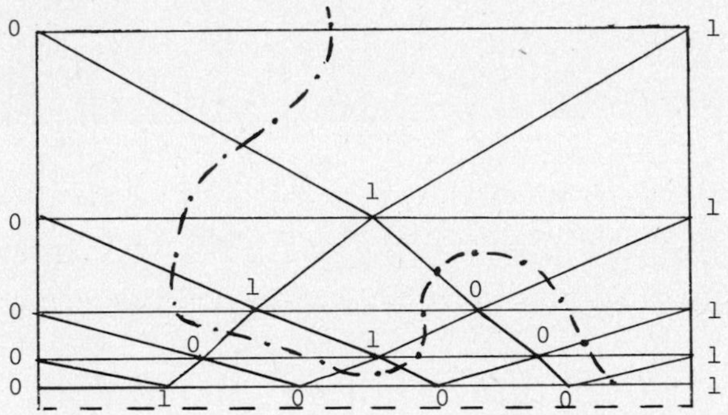

To eliminate advantage (1), we can use A(c-x) as the artificial label, with A an approximation to the Jacobian of f minus the identity (see VIII.5 and VIII.7.5). The second supposed "advantage" has been discouraging in computational experience--generally, Merrill's algorithm has the advantage when the homotopy algorithms "yoyo".

The advantage of Merrill's algorithm is its greater flexibility. The grid size can be reduced by any factor, as compared to $\frac{1}{2}$ for presently known triangulations with crogs. Also, one can make more use of the approximate inverse Jacobian in taking quasi-Newton steps with the option to restart the algorithm if convergence is poor. Of course, one can use a mixed strategy, and restart a homotopy algorithm as often as desired. Saigal, Solow, and Wolsey have a flexible algorithm [55] incorporating advantage (1) of homotopy algorithms with the flexibility of Merrill's approach.

CHAPTER X: TRIANGULATIONS WITH CONTINUOUS REFINEMENT OF GRID SIZE

X.1 Discussion. Triangulations with crogs were first introduced by Eaves [14] and Eaves and Saigal [16]. These triangulations, known as K_1, K_2 and K_3, differ from our triangulations of the same name. Diagrams of Eaves' K_1 and K_2 are in IX.1.3.

Recall the following:

1.1 Definition. G is a triangulation with crogs if for some sequence $\delta_k \downarrow 0$,

(i) G triangulates $(0,1] \times R^n$ and $G^0 \subseteq \bigcup_{k=0}^{\infty} R^n(k)$;

(ii) $G_k = \{\sigma \in G^n \mid \sigma \subseteq R^n(k)\}$ triangulates $R^n(k)$ for $k = 0,1,\ldots$; and

(iii) $\sigma \subseteq (0,2^{-k}] \times R^n$ implies $\text{mesh}_2'\sigma \leq \delta_k$ for $k = 0,1,\ldots$.

A natural way to construct such triangulations is to choose G_k to be $\varepsilon_k K_1$ or $\varepsilon_k J_1$ with $\varepsilon_k = \delta_k/\sqrt{n}$ and then join up the G_k's suitably. The first question is to choose the ε_k. Nobody knows yet how to construct triangulations with crogs where $\varepsilon_{k+1}/\varepsilon_k < \frac{1}{2}$ (except for, say, $n = 1$, when any ratio is possible.) However, even if one could construct such triangulations, one would incur some vertical paths between levels of lengths greater than $n+1$. The importance of this result is that with an affine function (or with a continuously differentiable function and sufficiently small mesh) a homotopy algorithm follows a vertical path.

1.2 Lemma. Let G be a triangulation with crogs so that $\text{mesh}_2'G_{k+1}/\text{mesh}_2'G_k < \frac{1}{2}$ for some k. Then there exists $x \in R^n$ such that $[2^{-k-1},2^{-k}] \times \{x\}$ intersects more than $n+1$ simplices of G.

Proof. Let σ be a simplex of G_k with $\text{diam}_2'\sigma > 2\,\text{mesh}_2'G_{k+1}$, and let x and z be two points of σ with $\|x-z\|_2 > 2\,\text{mesh}_2'G_{k+1}$. Let $y \in R^n(k+1)$ be the vertex of a simplex ρ of G with σ as a facet. Then either $\|p(x) - p(y)\|_2$ or $\|p(z) - p(y)\|_2$ is greater than $\text{mesh}_2'G_{k+1}$; assume the first. Now the sequence of simplices met by $[2^{-k-1},2^{-k}] \times \{x\}$ starts with a simplex with $n+1$ vertices in $R^n(k+1)$ with none equal to y and ends with a simplex with y the only vertex in $R^n(k+1)$. Since each pivot transfers at most one vertex from $R^n(k+1)$ to $R^n(k)$, at least $n+2$ simplices are met. □

It would be useful to find a strengthening of 1.2 that related the maximum vertical path or the average vertical path to the ratio of meshes.

We will choose $\varepsilon_k = 2^{-k}$, so that $\delta_k = 2^{-k}\sqrt{n}$. To gain an idea of how the layers can be joined, we consider the case $n = 2$ and two successive levels, say $R^2(1)$ and $R^2(0)$. We then have

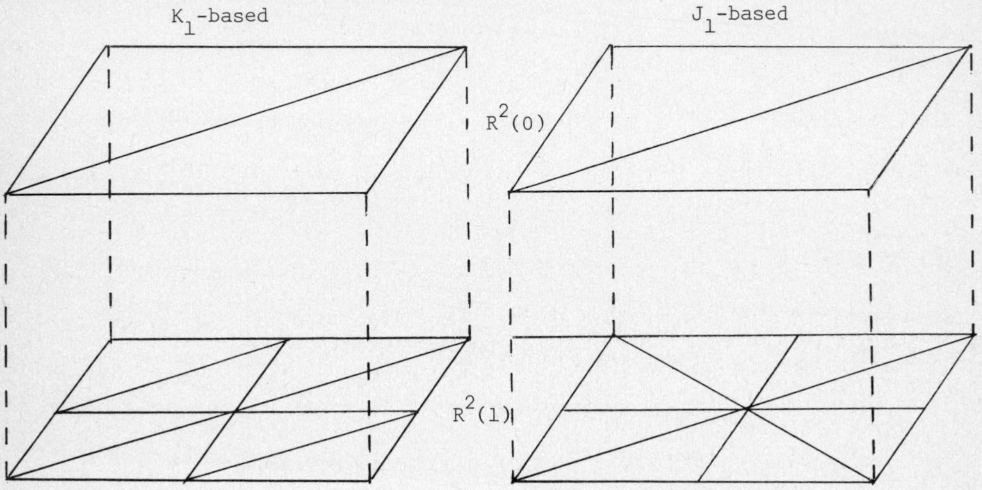

To triangulate $[\frac{1}{2},1] \times [0,1]^2$ we need two tetrahedra including the large triangles on top--the obvious fourth vertex is the central vertex in $R^2(1)$. We also need eight tetrahedra including the small triangles on the bottom. For each triangle there is a natural fourth vertex on the top. If these tetrahedra are removed, both cubes are left with four tetrahedra--a typical one has vertices at $(\frac{1}{2},\frac{1}{2},\frac{1}{2})^T$, $(\frac{1}{2},1,\frac{1}{2})^T$, $(1,1,0)^T$, and $(1,1,1)^T$. From this example we obtain the general method to construct (n+1)-simplices.

Suppose the grid size is ε on the bottom and 2ε on the top. Then the procedure is:

K-based simplex	J_1-based simplex
Pick $y = y^{-1}$ on the bottom.	Pick $y = y^{-1}$ on the bottom.
Pick a permutation π of N_0. Let $\pi(j) = 0$.	Pick a permutation π of N_0. Let $\pi(j) = 0$.
	Pick a sign vector $s \in R^n$.
Set $y^i = y^{i-1} + \delta v^{\pi(i)}$, $0 \leq i < j$.	Set $y^i = y^{i-1} + \delta s_{\pi(i)} v^{\pi(i)}$, $0 \leq i < j$.
Obtain y^j somehow from y^{j-1}.	Obtain y^j somehow from y^{j-1}.
Set $y^k = y^{k-1} + 2\delta v^{\pi(k)}$, $j < k \leq n$.	Set $y^k = y^{k-1} + 2\delta s_{\pi(k)} v^{\pi(k)}$, $j < k \leq n$.

The vertices $y^{-1},\ldots,y^{j-1}$ lie on the bottom level, while $y^j,\ldots,y^n$ lie on the top. The first $j+1$ vertices form a j-simplex of δK (δJ_1) on the bottom, and the rest form an (n-j)-simplex of $2\delta K$ $(2\delta J_1)$ on the top.

X.2 The Triangulation J_3 [64]. We will use a grid size of 1 on $R^n(0)$ and hence of 2^{-k} on $R^n(k)$. Thus the grid size equals the 0th coordinate.

2.1 Definition. The set of vertices of J_3 is $J_3^0 = \{y \in R^{n+1} | y_0 = 2^{-k}$ for some integer $k \geq 0$, y_i/y_0 is integral for $i \in N\}$. The central vertices of J_3 form a set $J_3^{0c} = \{y \in J_3^0 | y_i/y_0$ is odd for $i \in N\}$. Define $w: J_3^{0c} \to R^{n+1}$ by setting $w_i(y)$ equal to -1 or +1 according to whether y_i/y_0 is 1 or 3 modulo 4.

Let $y \in J_3^{0c}$ with $y_0 \leq \frac{1}{2}$ and consider $z = y - y_0 w(y)$. We have $z_0 = y_0 - y_0 w_0(y) = 2y_0$ and, for each $i \in N$, $z_i/y_0 = y_i/y_0 - w_i(y) \equiv 2 \bmod 4$; hence z_i/z_0 is an odd integer. Thus $z \in J_3^{0c}$ and z is the closest central vertex to y in the level above y. Now let $x = y + y_0 v^0$ be the point directly above y in the next-higher level. Then $x-z = (0, y_0 w(y)^T)^T$. Thus if $p(x) \in 2y_0 \bar{j}_1(z', \pi', t)$, we must have $z' = p(z)$, π arbitrary and $t = (w_1(y),\ldots,w_n(y))^T$. This suggests that to keep our simplices small we should ensure that $s_{\pi(k)} = w_{\pi(k)}(y)$ for $j < k \leq n$.

The remaining question in our general procedure is how to obtain y^j from y^{j-1}. Note that $y^{j-1} = y + y_0 \sum_{i=0}^{j-1} s_{\pi(i)} v^{\pi(i)}$. Hence for $h = \pi(i)$, $0 \leq i < j$, y_h^{j-1}/y_0 is even and we can set $y_h^j = y_h^{j-1}$. For $h = \pi(k)$, $j < k \leq n$, we have y_h^{j-1}/y_0 odd. We could set $y_h^j = y_h^{j-1} \pm y_0$; but since $y_h^n = y_h^j + 2y_0 w_h(y)$, it is natural to choose $y_h^j = y_h^{j-1} - y_0 w_h(y)$.

2.2 Definition. Let $y \in J_3^{0c}$ with $y_0 \leq \frac{1}{2}$, π be a permutation of N_0 with $\pi(j) = 0$, and $s \in R^n$ be a sign vector. Let $w = w(y)$. Then $j_3(y,\pi,s)$ is the simplex $\sigma = \langle y^{-1},\ldots,y^n \rangle$, where

$$y^{-1} = y;$$

$$y^i = y^{i-1} + y_0 s_{\pi(i)} v^{\pi(i)}, \quad 0 \leq i < j;$$

$$y^j = y^{j-1} - y_0 \textstyle\sum_{\ell=j}^n w_{\pi(\ell)} v^{\pi(\ell)}; \quad \text{and}$$

$$y^k = y^{k-1} + 2y_0 w_{\pi(k)} v^{\pi(k)}, \quad j < k \leq n.$$

Finally, J_3 is the collection of all such $j_3(y,\pi,s)$. (Note that σ is independent of $s_{\pi(k)}$, $j < k \leq n$. We generally indicate these coordinates of s by dots.)

Before proving that J_3 is indeed a triangulation with crogs, we illustrate how it triangulates $(0,1] \times R^n$ for $n = 1,2$.

2.3 Examples.

n = 1

Central vertices are indicated by heavy dots.

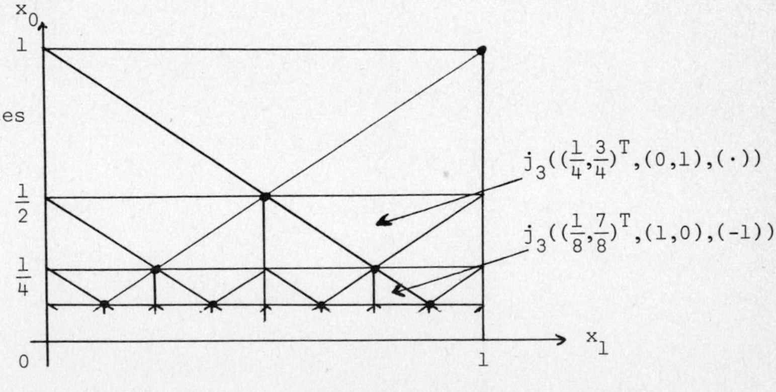

n = 2 We show the triangulation of $[\frac{1}{2},1] \times [0,1]^2$. The triangulation of $(0,1] \times R^2$ is obtained by rotations, replications, and shrinking operations. Indeed, the triangulation of $[\frac{1}{2},1] \times [0,2]^2$ is obtained by rotations around EK; that of $[\frac{1}{2},1] \times R^2$ (say $\tilde{J}_3$) by replication. Then $J_3 = \tilde{J}_3 \cup \frac{1}{2}\tilde{J}_3 \cup \ldots$, with $2^{-k}\tilde{J}_3$ triangulating $[2^{-k-1},2^{-k}] \times R^2$.

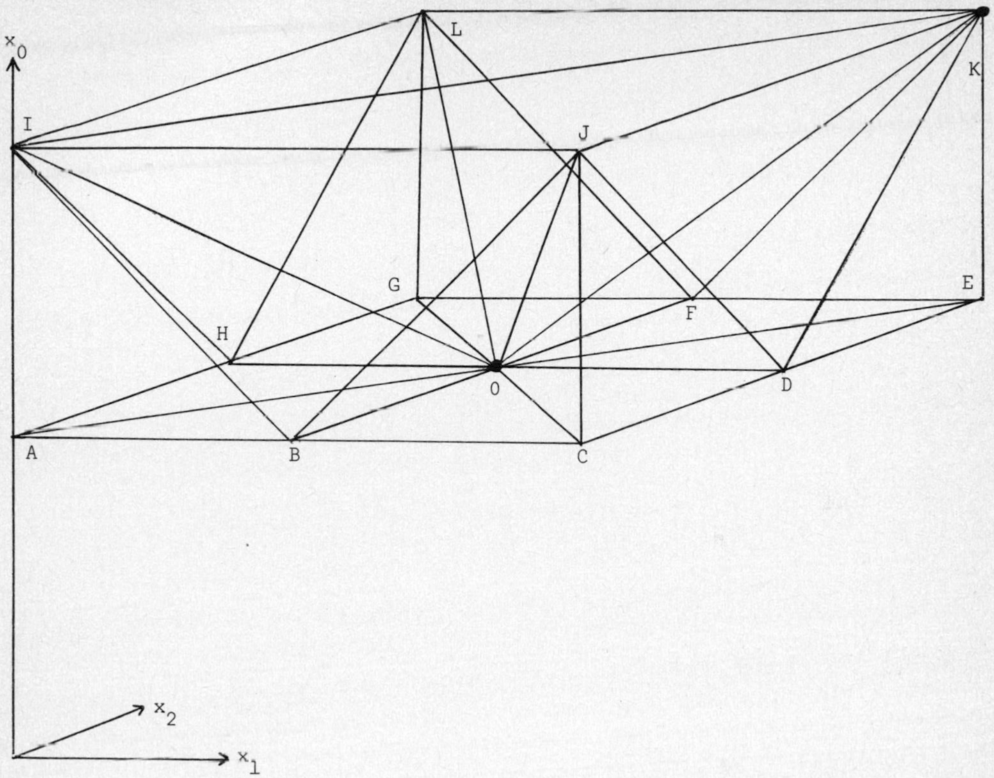

A list of the simplices follows. For each simplex $y^{-1} = O = (\frac{1}{2},\frac{1}{2},\frac{1}{2})^T$ with $w(y^{-1}) = (-1,-1,-1)^T$.

Simplex	y^{-1}	π	s
⟨OBAI⟩	O	(2,1,0)	(-1,-1)
⟨OBCI⟩	O	(2,1,0)	(+1,-1)
⟨ODCI⟩	O	(1,2,0)	(+1,-1)
⟨ODEI⟩	O	(1,2,0)	(+1,+1)
⟨OFEI⟩	O	(2,1,0)	(+1,+1)
⟨OFGI⟩	O	(2,1,0)	(-1,+1)
⟨OHGI⟩	O	(1,2,0)	(-1,+1)
⟨OHAI⟩	O	(1,2,0)	(-1,-1)
⟨OBJI⟩	O	(2,0,1)	(·,-1)
⟨ODKJ⟩	O	(1,0,2)	(+1,·)
⟨OFKL⟩	O	(2,0,1)	(·,+1)
⟨OHLI⟩	O	(1,0,2)	(-1,·)
⟨OKJI⟩	O	(0,2,1)	(·,·)
⟨OKLI⟩	O	(0,1,2)	(·,·)

2.4 Theorem. J_3 is a triangulation with continuous refinement of grid size, with corresponding $\{\delta_k\} = \{2^{-k}\sqrt{n}\}$.

Proof. We show first that J_3 satisfies 1.1(ii) and (iii), assuming that it triangulates $(0,1] \times R^n$. With this assumption, clearly no two faces of simplices in J_{3k} intersect--we must prove that they cover $R^n(k)$. Since δJ_1 triangulates R^n, we can project $2^{-k}J_1$ onto a triangulation of $R^n(k)$. Let $\tau = j_1(z,\pi',t)$ be a simplex of $2^{-k}J_1$. Define y by $y_0 = 2^{-k-1}$, $p(y) = z + y_0 t$. Then $y \in J_3^{0c}$ with $w(y) = t$. Let $\pi = (0,\pi'(1),\ldots,\pi'(n))$ and s be an arbitrary sign vector in R^n. The simplex $\sigma = j_3(y,\pi,s)$ then has as a face the projection ρ of τ. Hence $\rho \in J_{3k}$, and since τ was arbitrary, the simplices of J_{3k} and their faces cover $R^n(k)$.

For 1.2(iii), consider $\sigma \subseteq (0,2^{-k}] \times R^n$; then $\sigma = j_3(y,\pi,s)$ with $y_0 \leq 2^{-k-1}$ and $|y_i^{\ell} - y_i^m| \leq 2y_0 \leq 2^{-k}$ for $-1 \leq \ell,m \leq n$ and $i \in N$. Hence $\text{mesh}_2'\sigma \leq 2^{-k}\sqrt{n} = \delta_k$.

It remains to show that J_3 is a triangulation. Note that J_3 is locally finite. Thus we must prove that every $x \in (0,1] \times R^n$ lies in the closure of some $\sigma = j_3(y,\pi,s)$, and that the vertices of σ in the carrier of x in $\bar{\sigma}$ are uniquely determined from x.

Let $2^{-k} < x^0 \leq 2^{-k+1}$ and choose $y \in J_3^{0c} \cap R^n(k)$ as follows. We set $y_0 = 2^{-k}$ and y_i to be a closest odd multiple of y_0 to x_i for each $i \in N$. Then $y_0 \geq x_0 - y_0 > 0$ and $|x_i - y_i| \leq y_0$ for $i \in N$. With $w = w(y)$, for each $i \in N$ we have $y_0 \geq |x_i - y_i| > x_0 - y_0$ or $x_0 - y_0 \geq w_i(x_i - y_i) \geq y_0 - x_0$. Hence we can find a permutation π of N_0 (with $\pi(j) = 0$, say) and a sign vector $s \in R^n$ so that

$$\begin{aligned} y_0 &\geq s_{\pi(0)}(x_{\pi(0)} - y_{\pi(0)}) \geq \cdots \geq s_{\pi(j-1)}(x_{\pi(j-1)} - y_{\pi(j-1)}) \geq x_0 - y_0 \\ &\geq w_{\pi(j+1)}(x_{\pi(j+1)} - y_{\pi(j+1)}) \geq \cdots \geq w_{\pi(n)}(x_{\pi(n)} - y_{\pi(n)}) \geq y_0 - x_0. \end{aligned} \qquad (*)$$

Denote the terms above $\alpha_{-1} = y_0, \alpha_0, \ldots, \alpha_{j-1}, \alpha_j = x_0 - y_0, \alpha_{j+1}, \ldots, \alpha_n, \alpha_{n+1} = y_0 - x_0$. Define β by $\beta_i = (\alpha_i - \alpha_{i+1})/y_0$ for $-1 \leq i < j$ and $\beta_k = (\alpha_k - \alpha_{k+1})/2y_0$ for $j \leq k \leq n$. Then $\beta \geq 0$ and $\sum_{i=-1}^{i=n} \beta_i = \sum_{i=-1}^{j-1} \beta_i + \sum_{k=j}^{n} \beta_k = (\alpha_{-1} - \alpha_j)/y_0 +$

$(\alpha_j-\alpha_{n+1})/2y_0 = 1 - (x_0-y_0)/y_0 + (x_0-y_0)/2y_0 - (y_0-x_0)/2y_0 = 1.$

Let $\sigma = j_3(y,\pi,s) = \langle y^{-1},\ldots,y^n\rangle$ and consider $z = \sum_{i=-1}^{n} \beta_i y^i \in \bar{\sigma}$. We then have

$$\begin{aligned} z &= y^{-1} + \textstyle\sum_{i=0}^{j-1} \beta_i(y^i-y^{-1}) + \beta_j(y^j-y^{-1}) + \sum_{k=j+1}^{n} \beta_k(y^k-y^{-1}) \\ &= y + \textstyle\sum_{i=0}^{j-1} \beta_i(y^i-y^{-1}) + (\sum_{m=j}^{n} \beta_m)(y^j-y^{-1}) + \sum_{k=j+1}^{n} \beta_k(y^k-y^j) \\ &= y + \textstyle\sum_{i=0}^{j-1} \beta_i(y_0 \sum_{\ell=0}^{i} s_{\pi(\ell)}v^{\pi(\ell)}) \\ &\qquad + (\textstyle\sum_{m=j}^{n} \beta_m)(y_0 \sum_{\ell=0}^{j-1} s_{\pi(\ell)}v^{\pi(\ell)} - y_0 \sum_{\ell=j}^{n} w_{\pi(\ell)}v^{\pi(\ell)}) \\ &\qquad + \textstyle\sum_{k=j+1}^{n} \beta_k(2y_0 \sum_{\ell=j+1}^{n} w_{\pi(\ell)}v^{\pi(\ell)}) \\ &= y + \textstyle\sum_{\ell=0}^{j-1} s_{\pi(\ell)}v^{\pi(\ell)}(y_0 \sum_{i=\ell}^{j-1} \beta_i + y_0 \sum_{m=j}^{n} \beta_m) + v^{\pi(j)}(y_0 \sum_{m=j}^{n} \beta_m) \\ &\qquad + \textstyle\sum_{\ell=j+1}^{n} w_{\pi(\ell)}v^{\pi(\ell)}(2y_0 \sum_{k=\ell}^{n} \beta_k - y_0 \sum_{m=j}^{n} \beta_m). \end{aligned}$$

Now note that $y_0 \sum_{i=\ell}^{n} \beta_i = \alpha_\ell$ for $-1 \le \ell \le j$, and $2y_0 \sum_{k=\ell}^{n} \beta_k - y_0 \sum_{m=j}^{n} \beta_m = 2y_0(\alpha_\ell-\alpha_{n+1})/2y_0 - y_0(\alpha_j-\alpha_{n+1})/2y_0 = \alpha_\ell$ for $j < \ell \le n$. Hence

$$z = y + \sum_{\ell=0}^{j-1} \alpha_\ell s_{\pi(\ell)}v^{\pi(\ell)} + \alpha_j v^{\pi(j)} + \sum_{\ell=j+1}^{n} \alpha_\ell w_{\pi(\ell)}v^{\pi(\ell)},$$

and from the definition of α, $z = y + x-y = x$. We have established that each $x \in (0,1]\times R^n$ lies in the closure of some $\sigma \in J_3$.

Now let $x \in \bar{\sigma}$ with $\sigma = \langle y^{-1},\ldots,y^n\rangle = j_3(y,\pi,s) \in J_3$, and let $\{y^\ell, \ell \in L\}$ be the set of vertices of the carrier of x in $\bar{\sigma}$. As in III.3.2, $L = \{\ell|\beta_\ell > 0\}$ where β is as above. For every $z \in \bar{\sigma}$ we can find a unique $\beta(z)$ so that $z = \sum_{i=-1}^{n} \beta_i(z)y^i$. From this z we can obtain $\alpha(z)$ by $\alpha_\ell(z) = y_0 \sum_{i=\ell}^{n} \beta_i(z)$ for $-1 \le \ell \le j$ and $\alpha_\ell(z) = 2y_0 \sum_{k=\ell}^{n} \beta_k(z) - y_0 \sum_{m=j}^{n} \beta_m(z)$ for $j < \ell \le n$. Further, the terms of $\alpha(z)$ are as in (*); that is, $\alpha_{-1}(z) = y_0$, $\alpha_j(z) = -\alpha_{n+1}(z) = z_0-y_0$, $\alpha_\ell(z) = s_{\pi(\ell)}(z_{\pi(\ell)}-y_{\pi(\ell)})$ for $0 \le \ell < j$ and $\alpha_k(z) = w_{\pi(k)}(z_{\pi(k)}-y_{\pi(k)})$ for $j < k \le n$. The proof of this is similar to that in III.3.2.

For $\ell \in L$, $\beta(y^\ell)$ has the form $(0,\ldots,0,1,0,\ldots,0)$ and $\alpha(y^\ell)$ the form

$(y_0,\ldots,y_0,0,\ldots,0)$ if $\ell < j$ and $(y_0,\ldots,y_0,-y_0,\ldots,-y_0)$ if $\ell \geq j$. In both cases the last y_0 is in position ℓ and there is a strict inequality between positions ℓ and $\ell+1$ in $\alpha(x)$.

We can use these "α-vectors" to construct the y^ℓ, $\ell \in L$, directly from x. We distinguish two cases.

<u>Case 1</u>. $2^{-k} < x_0 < 2^{-k+1}$. Then y_0 is uniquely determined from x, and $y_0 > x_0-y_0 > 0 > y_0-x_0 > -y_0$. Since $\pm(x_i-y_i)$ must lie in $[y_0-x_0,y_0]$ for each $i \in N$, it is easily checked that y_i is uniquely determined from x unless $\pm(x_i-y_i) = y_0$. In this case x_i is an even multiple of y_0 and y_i can be either adjacent odd multiple; further, each y^ℓ, $\ell \in L$, has ith coordinate x_i.

We now use the description of $\alpha(y^\ell)$ for $\ell \in L$ to show how all y^ℓ, $\ell \in L$, can be obtained. Pick $\gamma \in (y_0-x_0,y_0]$; there will be one y^ℓ, denoted $z = z(\gamma)$, for each nontrivial interval between adjacent coordinates of $\alpha(x)$. If $\gamma > x_0-y_0$, z is obtained as follows. Round to the nearest even multiple of y_0 those coordinates of x at least γ from the nearest odd multiple of y_0. Round all other coordinates (including the 0th) to the nearest odd multiple of y_0. On the other hand, if $\gamma \leq x_0-y_0$, first round to the nearest even multiple of y_0 those coordinates (including the 0th) at least x_0-y_0 from the nearest odd multiple of y_0. For each remaining coordinate i, since $x_0-y_0 < y_0$, the nearest odd multiple y_i of y_0 is uniquely determined by x, and hence so is $w_i = w_i(y)$. Let $z_i = y_i + w_i$ if $w_i(x_i-y_i) > \gamma$ and $z_i = y_i - w_i$ if $w_i(x_i-y_i) \leq \gamma$.

Finally, notice that the rules above give all the vertices of the carrier of x in $\bar{\sigma}$ in terms of x alone; hence the same vertices are generated for the carrier of x in $\bar{\sigma}'$ for any $\sigma' \in J_3$ with $x \in \bar{\sigma}'$.

<u>Case 2</u>. $x_0 = 2^{-k+1}$. It is possible to proceed as above to generate the vertices of the carrier of x in $\bar{\sigma}$ and show them independent of y, π, and s. If $k \geq 1$, we also have to show that these are the same as the vertices of the carrier of x in $\bar{\sigma}'$ with $\sigma' = j_3(y',\pi',s')$ and $y' \in R^n(k-1)$. A simpler proof is obtained from the fact that J_1 triangulates R^n.

If $x \in \bar{\sigma}'' \cap R^n(k-1)$ for $\sigma'' = j_3(y'',\pi'',s'')$, $y'' \in R^n(k-1) \cup R^n(k)$, we find x lies in a face of a simplex $\tau = \langle z^0,\ldots,z^m\rangle \subseteq R^n(k-1)$ with τ a face of σ''.

The vertices of τ are the first $j''+1$ or the last $n-j''+1$ of those of σ'', with $\pi''(j'') = 0$. Now note that $z^i = z^{i-1} + 2^{-k+1} t_{\rho(i)} v^{\rho(i)}$ for $1 \leq i \leq m$, with $t \in R^n$ a sign vector and ρ a permutation of N. Thus τ is a face of a simplex of $2^{-k+1}J_1$, projected into $R^n(k-1)$. Since no two distinct faces of simplices of $2^{-k+1}J_1$ intersect, x can lie in only one possible face of a simplex of J_3. This completes the proof of case 2, and hence of the theorem. □

2.5 Example ($n = 3$). Let $x = (\frac{3}{8},\frac{1}{2},-\frac{5}{4},\frac{5}{8})^T$. We have $\frac{1}{4} < x_0 < \frac{1}{2}$. There are two choices for y: $(\frac{1}{4},\frac{1}{4},-\frac{5}{4},\frac{3}{4})^T$ or $(\frac{1}{4},\frac{3}{4},-\frac{5}{4},\frac{3}{4})^T$. If we pick the first, then $w(y) = (-1,-1,+1,+1)^T$. We have $y_0 = \frac{1}{4}$, $x_0 - y_0 = \frac{1}{8}$, $x_1 - y_1 = \frac{1}{4}$, $x_2 - y_2 = 0$, and $x_3 - y_3 = -\frac{1}{8}$. We can therefore write the string of inequalities (*) of the proof above in two ways:

$$y_0 \geq +1\cdot(x_1-y_1) \geq -1\cdot(x_3-y_3) \geq x_0-y_0 \geq +1\cdot(x_2-y_2) \geq y_0-x_0$$

$$\text{or } \frac{1}{4} \geq \frac{1}{4} \geq \frac{1}{8} \geq \frac{1}{8} \geq 0 \geq -\frac{1}{8},$$

with $\pi = (1,3,0,2)$, $s = (+1,\cdot,-1)$; or

$$y_0 \geq +1\cdot(x_1-y_1) \geq x_0-y_0 \geq +1\cdot(x_2-y_2) \geq +1\cdot(x_3-y_3) \geq y_0-x_0$$

$$\text{or } \frac{1}{4} \geq \frac{1}{4} \geq \frac{1}{8} \geq 0 \geq -\frac{1}{8} \geq -\frac{1}{8},$$

with $\pi = (1,0,2,3)$, $s = (+1,\cdot,\cdot)$.

Again choose the first. Let $\sigma = j_3(y,\pi,s) = \langle y^{-1},y^0,y^1,y^2,y^3\rangle$. We have $y^{-1} = (\frac{1}{4},\frac{1}{4},-\frac{5}{4},\frac{3}{4})^T$, $y^0 = (\frac{1}{4},\frac{1}{2},-\frac{5}{4},\frac{3}{4})^T$, $y^1 = (\frac{1}{4},\frac{1}{2},-\frac{5}{4},\frac{1}{2})^T$, $y^2 = (\frac{1}{2},\frac{1}{2},-\frac{3}{2},\frac{1}{2})^T$, and $y^3 = (\frac{1}{2},\frac{1}{2},-1,\frac{1}{2})^T$. From the string of inequalities above, $\alpha_{-1} = \frac{1}{4}$, $\alpha_0 = \frac{1}{4}$, $\alpha_1 = \frac{1}{8}$, $\alpha_2 = \frac{1}{8}$, $\alpha_3 = 0$, and $\alpha_4 = -\frac{1}{8}$; hence $\beta_{-1} = (\frac{1}{4} - \frac{1}{4})/\frac{1}{4} = 0$, $\beta_0 = (\frac{1}{4} - \frac{1}{8})/\frac{1}{4} = \frac{1}{2}$, $\beta_1 = (\frac{1}{8} - \frac{1}{8})/\frac{1}{4} = 0$, $\beta_2 = (\frac{1}{8} - 0)/(2\cdot\frac{1}{4}) = \frac{1}{4}$, and $\beta_3 = (0 - (-\frac{1}{8}))/(2\cdot\frac{1}{4}) = \frac{1}{4}$. One can check that $x = \sum \beta_i y^i = \frac{1}{2}y^0 + \frac{1}{4}y^2 + \frac{1}{4}y^3$ and that the same vertices occur if another choice of y, π, or s is made.

Alternatively, we can generate the vertices of the carrier of x directly by picking $\gamma \in (-\frac{1}{8},\frac{1}{4}]$. We find $z(\gamma) = (\frac{1}{4},\frac{1}{2},-\frac{5}{4},\frac{3}{4})^T = y^0$ for $\frac{1}{4} \geq \gamma > \frac{1}{8}$,

$z(\gamma) = (\frac{1}{2},\frac{1}{2},-\frac{3}{2},\frac{1}{2})^T = y^2$ for $\frac{1}{8} \geq \gamma > 0$, and $z(\gamma) = (\frac{1}{2},\frac{1}{2},-1,\frac{1}{2})^T = y^3$ for $0 \geq \gamma > -\frac{1}{8}$.

X.3 The Triangulation K_3. For consistency of notation we denote by K_3 the triangulation K_3' of [65]; it is closely related to Eaves' and Saigal's K_3 [16]. We discuss this relationship in [65].

The vertices of K_3 are the same as those of J_3. Central vertices are no longer of concern, but it is convenient to introduce a function t to mark which coordinates of a vertex are even multiples of the 0th coordinate.

3.1 Definition. $K_3^0 = \{y \in (0,1] \times R^n | y_0 = 2^{-k}$ for some integer $k \geq 0$, y_i/y_0 is integral for $i \in N\}$. Define $t: K_3^0 \to R^{n+1}$ by setting $t_i(y) = 0$ if y_i/y_0 is odd and $t_i(y) = 1$ otherwise. (Thus $t_0(y)$ always equals 0.)

Recall the general procedure of section 1. We have $y = y^{-1} \in K_3^0$ and a permutation π of N_0, with $\pi(j) = 0$. Then $y^i = y^{i-1} + y_0 v^{\pi(i)}$ for $0 \leq i < j$ and $y^k = y^{k-1} + 2y_0 v^{\pi(k)}$ for $j < k \leq n$. We must specify how y^j is determined from y^{j-1}. Let $t = t(y)$ and note that $y^{j-1} = y + y_0 \sum_{i=0}^{j-1} v^{\pi(i)}$. Thus for $h \in N$, y_h^{j-1}/y_0 is odd if $t_h = 1$ and $\pi^{-1}(h) < j$ or if $t_h = 0$ and $\pi^{-1}(h) > j$, and even otherwise. To keep the simplex small we want $y_h^j = y_h^{j-1}$ or $y_h^{j-1} - y_0$. It is then natural to let

$$y^j = y^{j-1} + y_0(-\sum_{i=0}^{j-1} t_{\pi(i)} v^{\pi(i)} + v^0 - \sum_{k=j+1}^{n} (1-t_{\pi(k)}) v^{\pi(k)}),$$

so that $y^j \in K_3^0$. The only problem with setting K_3 equal to the collection of all such simplices $\langle y^{-1},\ldots,y^n \rangle$ is that the result is not a triangulation. The difficulty is illustrated in the case $n = 1$ below.

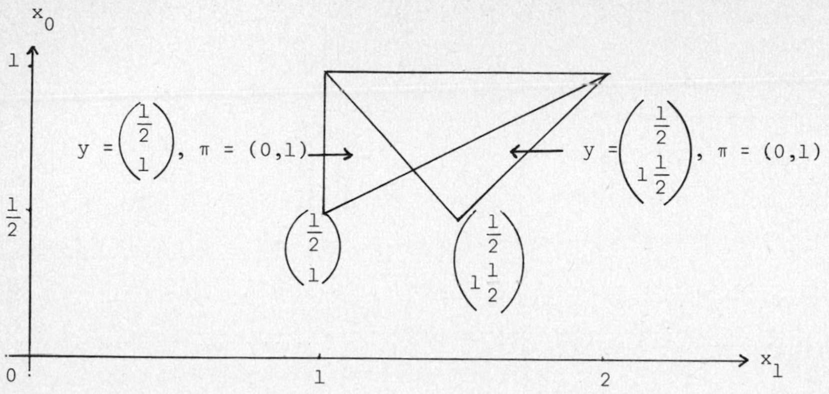

Note that two simplices intersect. We eliminate the problem by omitting some of the simplices.

3.2 Definition. Let $y \in K_3^0$ with $t = t(y)$ and let π be a permutation of N_0 with $\pi(j) = 0$. We call (y,π) admissible if $t_{\pi(k)} = 0$ for all $j < k \leq n$. In this case, $k_3(y,\pi)$ is the simplex $\langle y^{-1},\ldots,y^n\rangle$ where

$$y^{-1} = y;$$

$$y^i = y^{i-1} + y_0 v^{\pi(i)}, \quad 0 \leq i < j;$$

$$y^j = y^{j-1} + y_0\left(-\textstyle\sum_{i=0}^{j-1} t_{\pi(i)} v^{\pi(i)} + v^0 - \sum_{k=j+1}^{n} v^{\pi(k)}\right);$$

$$y^k = y^{k-1} + 2y_0 v^{\pi(k)}, \quad j < k \leq n.$$

Finally, $K_3 = \{k_3(y,\pi) \mid (y,\pi) \text{ admissible}\}$.

3.3 Examples.

n = 1

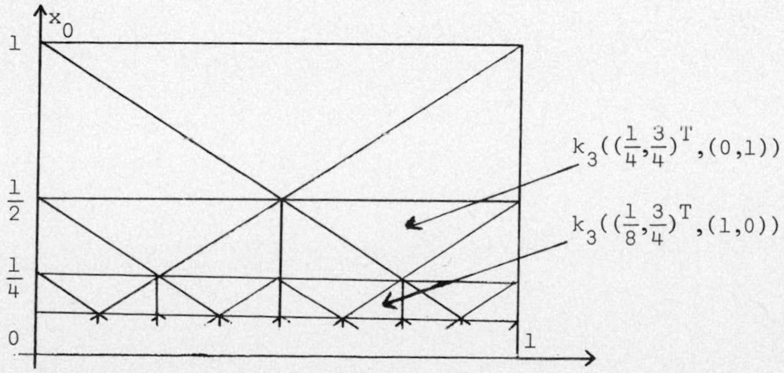

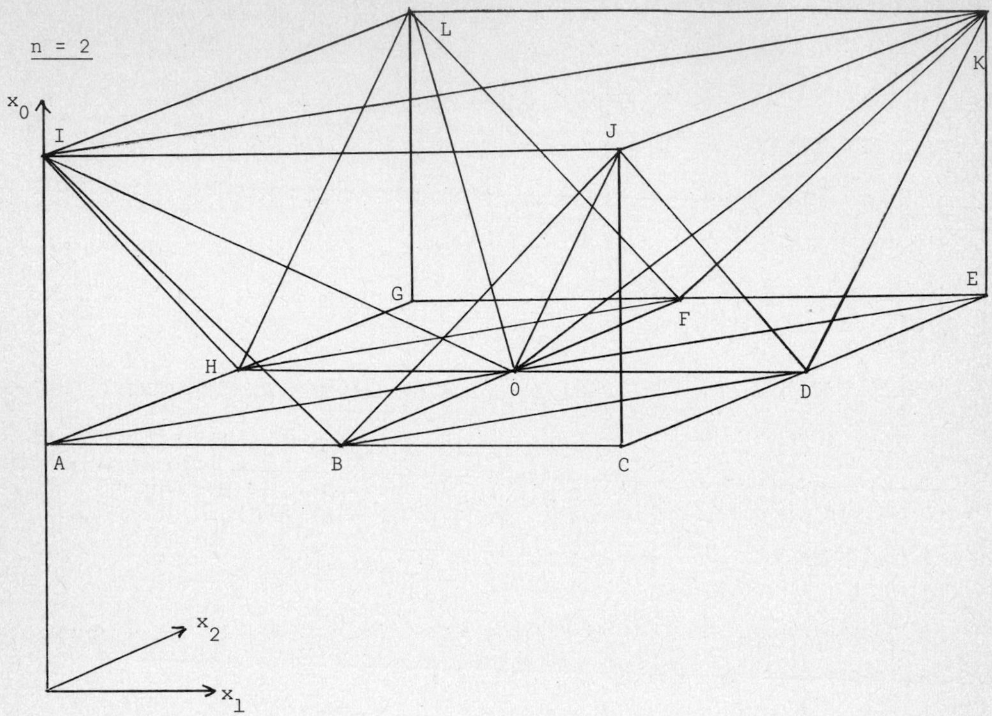

This is the triangulation by K_3 of $[\frac{1}{2},1] \times [0,1]^2$. A list of the simplices follows. (Note that without the stipulation of admissibility, other simplices intersecting these (such as AIJK, ABIL, BCJK) would appear.)

$k_3(y,\pi)$	y	π
⟨ABOI⟩	A	(1,2,0)
⟨BODJ⟩	B	(2,1,0)
⟨BCDJ⟩	B	(1,2,0)
⟨ODEK⟩	O	(1,2,0)
⟨OFEK⟩	O	(2,1,0)
⟨HOFL⟩	H	(1,2,0)
⟨HGFL⟩	H	(2,1,0)
⟨AHOI⟩	A	(2,1,0)
⟨BOIJ⟩	B	(2,0,1)
⟨ODJK⟩	O	(1,0,2)
⟨OFLK⟩	O	(2,0,1)
⟨HOIL⟩	H	(1,0,2)
⟨OIJK⟩	O	(0,1,2)
⟨OILK⟩	O	(0,2,1)

Note that $J_3 = K_3$ for $n = 1$, and J_3 and K_3 are very similar for $n = 2$.

For $n = 2$ we can observe the following very important difference between the two. Consider a vertical path near CJ in K_3. The following simplices are met: $\langle BCDJ \rangle$, $\langle BODJ \rangle$, $\langle BOIJ \rangle$ or $\langle ODJK \rangle$, and finally $\langle OIJK \rangle$. In fact a vertical path intersecting BCD or FGH meets 4 simplices of K_3; any vertical path intersecting ABDEFH meets only 3 simplices of K_3. With a starting point uniformly distributed in $[0,1]^2$, the expected number is $3\frac{1}{4}$. However, if the same analysis is applied to J_3, the number is only 3. While the difference seems small, it increases with n; for K_3 the general term is $n + 1 + \frac{1}{4}\binom{n}{2}$, while for J_3 it is $n+1$. A more complete analysis and comparison of triangulations form the next chapter.

<u>3.4 Theorem</u>. K_3 is a triangulation with continuous refinement of grid size.

<u>Proof</u>. See [65]. The proof follows that for J_3. For future reference we note that $\bar{k}_3(y,\pi)$ is the set of $x \in R^{n+1}$ with

$$y_0 \geq x_{\pi(0)} - y_{\pi(0)} + t_{\pi(0)}(x_0 - y_0) \geq \ldots \geq x_{\pi(n)} - y_{\pi(n)} + t_{\pi(n)}(x_0 - y_0) \geq y_0 - x_0. \quad (**)$$

X.4 The Pivot Rules of J_3 and K_3.

<u>4.1</u>. Let $\sigma' = j_3(y',\pi',s')$ and $\sigma = j_3(y,\pi,s)$ share the facet of σ opposite y^i, with $-1 \leq i \leq n$. For computational purposes we do not store either s or $w = w(y)$ but the composite vector $a = a(y,\pi) \in R^n$, with $a_{\pi(i)} = s_{\pi(i)}$ for $0 \leq i < j = \pi^{-1}(0)$ and $a_{\pi(k)} = w_{\pi(k)}$ for $j < k \leq n$. Clearly, σ can be recovered from y, π, and a. Let $a' = a(y',\pi')$. Table I shows how y', π' and a' can be obtained from y, π, a and i.

<u>4.2</u>. Let $\sigma' = k_3(y',\pi')$ and $\sigma = k_3(y,\pi)$ share the facet of σ opposite y^i, with $-1 \leq i \leq n$. It is convenient to store and update $t = t(y)$; let $t' = t(y')$. Table II shows how y', π' and t' can be obtained from y, π, t, and i. Again, let $j = \pi^{-1}(0)$.

		y'	π'	a'
$i = -1$	$j = 0$	$y - y_0(-1,a^T)^T$	$(\pi(1)\ldots\pi(n)\pi(0))$	a
	$j > 0$	$y + 2y_0 a_{\pi(0)} v^{\pi(0)}$	π	$a - 2a_{\pi(0)} u^{\pi(0)}$
$0 \leq i < j-1$		y	$(\pi(0)\ldots\pi(i+1)\pi(i)\ldots\pi(n))$	a
$i = j-1$	$a_{\pi(j-1)} = \nu_{\pi(j-1)}$	y	$(\pi(0)\ldots\pi(j)\pi(j-1)\ldots\pi(n))$	a
	$a_{\pi(j-1)} = -\nu_{\pi(j-1)}$	y	$(\pi(0)\ldots\pi(j-2),\pi(j)\ldots\pi(n)\pi(j-1))$	$a - 2a_{\pi(j-1)} u^{\pi(j-1)}$
$j \leq i < n$		y	$(\pi(0)\ldots\pi(i+1),\pi(i)\ldots\pi(n))$	a
$i = n$	$j = n$	$y + \frac{1}{2} y_0(-1,a^T)^T$	$(\pi(n),\pi(0),\ldots,\pi(n-1))$	a
	$j < n$	y	$(\pi(0)\ldots\pi(j-1),\pi(n),\pi(j)\ldots\pi(n-1))$	$a - 2a_{\pi(n)} u^{\pi(n)}$

Table I

			y'	π'	t'
$i = -1$	$j > 0$	$t_{\pi(0)} = 0$	$y + y_0 v^{\pi(0)}$	$(\pi(1)\ldots\pi(j-1),\pi(0),\pi(j)\ldots\pi(n))$	$t + v^{\pi(0)}$
		$t_{\pi(0)} = 1$	$y + y_0 v^{\pi(0)}$	$(\pi(1)\ldots\pi(n),\pi(0))$	$t - v^{\pi(0)}$
	$j = 0$		$y - y_0(-1,u^T)^T$	$(\pi(1)\ldots\pi(n),\pi(0))$	Recalculate
$0 \leq i < j-1$			y	$(\pi(0)\ldots\pi(i+1),\pi(i)\ldots\pi(n))$	t
$i = j-1$	$t_{\pi(j-1)} = 0$		y	$(\pi(0)\ldots\pi(j),\pi(j-1)\ldots\pi(n))$	t
	$t_{\pi(j-1)} = 1$		$y - y_0 v^{\pi(j-1)}$	$(\pi(j-1),\pi(0)\ldots\pi(j-2),\pi(j)\ldots\pi(n))$	$t - v^{\pi(j-1)}$
$j \leq i < n$			y	$(\pi(0)\ldots\pi(i+1),\pi(i)\ldots\pi(n))$	t
$i = n$	$j < n$		$y - y_0 v^{\pi(n)}$	$(\pi(n),\pi(0)\ldots\pi(n-1))$	$t + v^{\pi(n)}$
	$j = n$		$y + \frac{1}{2} y_0(-1,u^T)^T$	$(\pi(n),\pi(0)\ldots\pi(n-1))$	$(0,\ldots,0)$

Table II

CHAPTER XI: MEASURES OF EFFICIENCY FOR TRIANGULATIONS

Computational experience with the fixed-point algorithms of Chapters IV and VII-IX has shown a considerable sensitivity to the triangulation used. Our aim in this chapter is to define some theoretical measures that we hope will predict the relative efficiencies of different triangulations.

The algorithms we have discussed move from simplex to adjacent simplex, each such move requiring a natural or artificial function evaluation and usually a linear-programming pivot step. Thus it is natural to compare triangulations by comparing the length of various paths of adjacent simplices.

XI.1 The Triangulation H and a Crude Measure. Let us first introduce a triangulation H_1 of R^n. This triangulation has performed poorly in a number of computational tests compared to K_1 or J_1, and we want the measures we introduce to capture this difference.

Let $H_1^0 = Z^n = \{y \in R^n | y_i$ is integer for $i \in N\}$. Let P be the $n \times n$ matrix with -1's on the diagonal, +1's on the subdiagonal, and zeroes elsewhere. Then for every $y \in H_1^0$ and permutation π of N, $h_1(y,\pi)$ is the simplex $\langle y^0,\ldots,y^n \rangle$ with $y^0 = y$ and $y^i = y^{i-1} + p^{\pi(i)}$, $i \in N$, where p^j is the jth column of P. Let H_1 be the collection of all such $h_1(y,\pi)$. Then H_1 triangulates R^n-- indeed, the matrix P takes R^n into itself and transforms K_1 into H_1.

Note that $\bar{h}_1(y,\pi)$ is the set of $x \in R^n$ satisfying

$$1 \geq -\sum_{i=1}^{\pi(1)} (x_i - y_i) \geq \cdots \geq -\sum_{i=1}^{\pi(n)} (x_i - y_i) \geq 0.$$

For $n = 2,3$ we have the pictures:

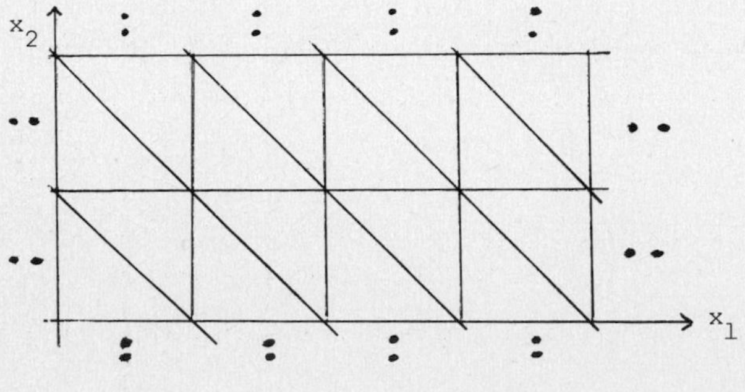

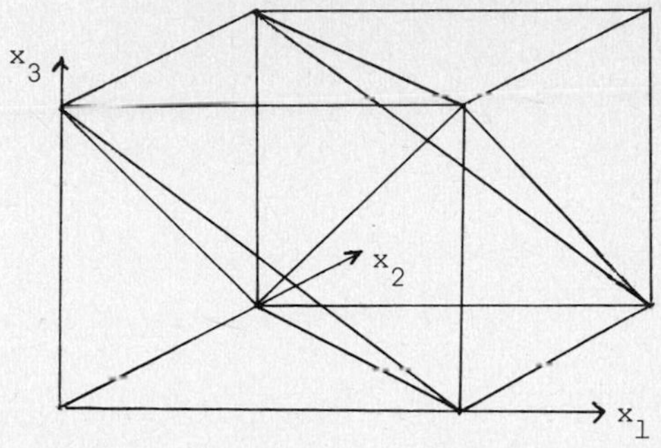

$\tilde{H}_1$ is obtained from H_1 as are $\tilde{K}_1$ and $\tilde{J}_1$ from K_1 and J_1. (See VIII.2.2.)

Throughout this chapter we shall call a facet of a triangulation G the closure of a member of G^{k-1}, where G consists of k-simplices. This convention makes many statements simpler--for example, we can easily give conditions for a point in R^n to lie in a facet of K_1, J_1, or H_1.

1.1 Lemma. $x \in R^n$ lies in a facet of

(a) K_1 iff $x_i \in Z$ for some $i \in N$ or $(x_i - x_j) \in Z$ for some $i \neq j \in N$;

(b) J_1 iff $x_i \in Z$ for some $i \in N$ or $(x_i - x_j)/2$ or $(x_i + x_j)/2 \in Z$ for some $i \neq j \in N$;

(c) H_1 iff $\sum_{k=i}^{j} x_k \in Z$ for some $i \leq j \in N$.

Proof. Trivial using the facetal descriptions of K_1, J_1, and H_1. □

It follows from 1.1 that K_1, J_1, and H_1 all triangulate the unit cube $C = \{x \in R^n | 0 \leq x \leq u\}$ in R^n. Thus one crude measure is the number of simplices contained in C. Unfortunately, this measure does not distinguish between K_1, J_1, and H_1; they all have $n!$ simplices in C. We can see this in two ways.

First, C has volume 1, while each $k_1(y,\pi)$, $j_1(y,\pi,s)$ and $h_1(y,\pi)$ has volume $1/n!$ Second, $k_1(y,\pi) \subseteq C \iff y = 0$, giving $n!$ simplices of K_1 in C. Similarly, $j_1(y,\pi,s) \subseteq C \iff y = u$, $s = -u$, giving $n!$ simplices of J_1 in C. With a little more work we can establish that $h_1(y,\pi) \subseteq C$ iff $y_1 = 1$ and $y_i = 0$ or 1, according to whether $i-1$ precedes i in the permutation π or not, for $i = 2,3,\ldots,n$. Thus there is one such simplex in C for each permutation π and n! in total.

The crude measure introduced above fails to distinguish between the three triangulations. However, there is much computational experience, notably that reported in Saigal [52], showing that H_1 is far inferior to K_1, especially for large n. In the next three sections we discuss more sophisticated measures.

The poor behavior of H_1 is rather unexpected. Since P takes the triangulation K_1 into H_1, finding a fixed point of f using H_1 is equivalent to finding a fixed point of $g = P^{-1}fP$ using K_1, in the sense that the same number of simplices will be met. One would expect that functions like g would occur as frequently as functions like f and that K_1 and H_1 would be comparable. However, computational experience with many functions and starting points overwhelmingly indicates the superiority of K_1.

Before rejecting the crude measure above, let us note an interesting and unsolved problem: what is the triangulation of C into the fewest simplices? For n = 3, 4, and 5 triangulations yielding 5, 16, and 68 simplices have been constructed by Mara [47], but there is no known way to extend the construction. For n = 3 one has the picture

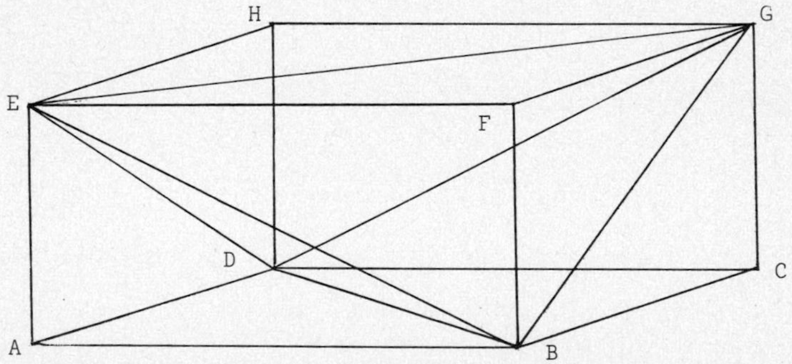

with simplices $\langle ABDE\rangle$, $\langle CBDG\rangle$, $\langle FBGE\rangle$, $\langle HDEG\rangle$ and $\langle BDEG\rangle$. Two comments are appropriate. Firstly, there is no easy way to describe the pivot rules of these triangulations for n = 4 or 5. Secondly, the advantage of these triangulations over K_1 or J_1 is not as great as it seems. If G is replaced by 2G, the number of simplices in C (an "n-dimensional" measure) is reduced by a factor of 2^n, whereas the number of simplices in a path between two points (a "1-dimensional" measure) is reduced only by a factor of 2. Since "1-dimensional" measures are closest to predicting the work of an algorithm, the advantage of Mara's triangulations is only of the order of $\sqrt[3]{5/6}$, $\sqrt[4]{16/24}$, and $\sqrt[5]{68/120}$, about 10% in each case.

XI.2 The Diameter of a Triangulation. Let G be a triangulation of R^n and suppose that $\tilde{G} = \{\sigma \in G | \sigma \subseteq C\}$ triangulates C and that $\tilde{G}$ can be replicated to give all of G, i.e., that $G = \bigcup_{y \in Z^n} \{\sigma + \{y\} | \sigma \in \tilde{G}\}$. The diameter measure was introduced by Saigal, Solow, and Wolsey [55] to compare such triangulations G by means of their restrictions $\tilde{G}$.

Let τ and τ' be two facets of $\tilde{G}$ in ∂C. Let $\sigma_0, \ldots, \sigma_m$ be a sequence of simplices of $\tilde{G}$ with σ_{i-1} and σ_i adjacent for $1 \leq i \leq m$. If τ is a facet of σ_0 and τ' a facet of σ_m, we call $\sigma_0, \ldots, \sigma_m$ a path from τ to τ' of length $m+1$. We define the distance between τ and τ' to be the minimum length of a path between them. The diameter of G is the maximum such distance.

To justify this measure, Saigal, Solow, and Wolsey proved the following result.

2.1 Lemma. Let $F_i(x) = \{x + A(c^i - x)\}$ for $i = 0,1$, with A an $n \times n$ nonsingular matrix. Let L be the line segment $\{(\lambda, ((1-\lambda)c^0 + \lambda c^1)^T)^T | \lambda \in [0,1]\}$. Let the special triangulation G be such that L meets no simplex of G^+ of dimension less than n. Suppose $(i, c^{iT})^T \in \tau_i \in G^n$ for $i = 0,1$. Then if Merrill's algorithm is used to compute the fixed point c^1 of F^1 using the triangulation G, the simplices generated are precisely those meeting L. If $G = \tilde{K}_1$, $\tilde{J}_1$, or $\tilde{H}_1$, the simplices generated form a minimum length path from τ_0 to τ_1.

Proof. The corresponding vector labelling ℓ is defined by $\ell(x) = A(x_0 c^1 + (1-x_0)c^0 - p(x))$. Thus $\ell(x) = 0$ iff $x \in L$. If L meets no simplex of G^+ of dimension less than n, every complete simplex is very complete and meets L. The first part follows.

Now let $G = \tilde{K}_1$. (The proof is similar for $\tilde{J}_1$ and $\tilde{H}_1$.) Then any path of simplices from τ_0 to τ_1 crosses the hyperplane $x_i = m$ for every integer m in (c_i^0, c_i^1) and the hyperplane $x_i - x_j = \ell$ for every integer ℓ in $(c_i^0 - c_j^0,\ c_i^1 - c_j^1)$, where if $\lambda > \mu$, (λ,μ) is interpreted as (μ,λ). By 1.1 each such crossing necessitates one pivot. But each hyperplane of this form is crossed just once by the path of simplices that intersect L, and hence the length of this path is minimal. $\square$

Saigal, Solow, and Wolsey found the diameter of K_1 to be $O(n^2)$. Saigal [52] subsequently showed that the diameter of H_1 was at least $O(n^3)$. Thus the diameter is one measure that discriminates between K_1 and H_1.

XI.3 The Directional Density. This section and the next are based on [65].

Lemma 2.1 motivates us to count the number of simplices meeting a line segment. For $x,d \in R^n$, $\lambda > 0$, let $[x, x+\lambda d]$ denote $\{x + \mu d | \mu \in [0,\lambda]\}$. Let G be a triangulation of R^n.

3.1 Definition. Let $N(G,x,d,\lambda)$ be the number of simplices of G intersecting $[x, x+\lambda d]$ divided by λ. Let $N(G,d,\lambda)$ be the limit as $\rho \to \infty$ (if it exists) of the average of $N(G,x,d,\lambda)$ for x uniformly distributed in $B(0,\rho)$. Let $N(G,d)$ be the limit as $\lambda \to \infty$ (if it exists) of $N(G,d,\lambda)$. Finally, $N(G)$ is the average of $N(G,d)$ for d uniformly distributed in ∂B^n.

We call $N(G,d)$ the directional density of G in direction d and $N(G)$ the average directional density. Note that $N(G,x,d,\lambda)$ is the number of simplices met per unit step size. To eliminate the effects of the starting point, we average over x lying in a large ball. For any replicable triangulation we can average over any replicable shape. To eliminate any effects on the ending point, we let λ tend to ∞. Finally, to get a measure independent of the direction d, we average over ∂B^n. Note that we could also average over $\{d \in R^n | \|d\|_\infty \leq 1\}$ or $\{d \in R^n | \sum_{i \in N} |d_i| \leq 1\}$. We chose the Euclidean norm because the averaging is easier. Any choice is related to the question of how one's guesses to a solution to a problem are distributed about the true solution. We presume a spherical distribution; the other sets are appropriate to a cubical or octahedral distribution.

For any triangulation G of R^n, G^{n-2} is a countable collection of sets of dimension $n-2$. Thus for almost all x, $\{x + \lambda d | \lambda \in R\}$ meets no simplex of G^+ of dimension less than $n-1$. If x and $x + \lambda d$ lie in n-simplices of G,

$[x, x+\lambda d]$ meets one more n-simplex of G than (n-1)-simplex of G. Thus the computation of $N(G,x,d,\lambda)$ can be made by counting the number of points of $[x, x+\lambda d]$ lying in facets of G, for almost all x.

From these observations it is easy to obtain

3.2 Theorem.

(a) $N(K_1,d) = \sum_i |d_i| + \sum_{i<j} |d_i-d_j|$

(b) $N(J_1,d) = \sum_i |d_i| + \sum_{i<j} \frac{1}{2}(|d_i+d_j| + |d_i-d_j|)$

(c) $N(H_1,d) = \sum_{i\leq j} |\sum_{k=i}^{j} d_k|$

(d) Let $g_n = 2\Gamma(n/2)/(n-1)\sqrt{\pi}\,\Gamma((n-1)/2)$. Then $N(K_1) = N(J_1) = (n + \sqrt{2}\binom{n}{2})g_n$ and $N(H_1) = \sum_i (n+1-i)\sqrt{i}\, g_n$.

[All indices of sums run over N.]

Proof. As observed above, we may average over x in any replicable shape, and we may count facets crossed rather than simplices met.

(a) As z traverses $[x, x+\lambda d]$, z_i goes from x_i to $x_i + \lambda d_i$ and $z_i - z_j$ from $x_i - x_j$ to $x_i - x_j + \lambda(d_i-d_j)$. From 1.1 the number of facets crossed is the sum over i of the number of integers in $[x_i, x_i + \lambda d_i]$ plus the sum over $i < j$ of the number of integers in $[x_i - x_j, x_i - x_j + \lambda(d_i-d_j)]$. As x_i and x_j vary uniformly in $[0,1]$, these numbers average $\lambda|d_i|$ and $\lambda|d_i - d_j|$, respectively. Part (a) follows.

Parts (b) and (c) are proved similarly.

(d) For each i the average of $|d_i|$ over ∂B^n is the same. By rather messy integration it turns out to be g_n. By a change of coordinates, the average value of $|d_i - d_j|$ is $\sqrt{2}\, g_n$ and that of $|\sum_{k=i}^{j} d_k|$ is $\sqrt{j-i+1}\, g_n$. Part (d) then follows from (a)-(c). □

Asymptotically we have $N(K_1) = N(J_1) \sim n^2 g_n/\sqrt{2}$, while $N(H_1) \sim 4n^{5/2} g_n/15$. Again the inferiority of H_1 is predicted. The directional density also indicates that K_1 is best when the direction d can be predicted (rotate K_1 so that $d = u$) and that J_1 is less sensitive to direction than K_1. ($N(J_1,d)$ is the average of $N(K_1,\bar{d})$ for all 2^n $\bar{d}$'s with $|\bar{d}_i| = |d_i|$, $i \in N$.)

The proof of 3.2 shows that the same minimal number of simplices is met by any sequence of simplices crossing the separating facets only once. For example, Wolsey [74] considered Merrill's algorithm using integer labels and the triangulation K_1, applied to the situation of 2.1. While Wolsey's proof contains errors, his conclusion is valid; with appropriate integer labelling, a minimum number of simplices is met.

We can extend the analysis above to take into consideration the fact that function evaluations are much more costly than artificial label evaluations. If the work required to compute a function evaluation dominates that to make a linear-programming pivot, these more complicated measures describe the behavior of a triangulation used in Merrill's algorithm more accurately than those above. Similarly, we can compare the triangulations K_1, J_1, and H_1 with the first layers of J_3 and K_3, suitably modified for Merrill's algorithm. It turns out that H_1 is again inefficient while K_1, J_1, J_3, and K_3 are comparable for function evaluations. However, J_3 and K_3 require approximately half as many artificial label evaluations, thus saving on linear-programming pivots. When Merrill's algorithm starts to converge fast, very few artificial labels are computed and the differences are negligible.

One can also compute measures for the triangulations $K_2(m)$ and $J_2(m)$ of S^n. The results indicate that it is preferable to project δK_1 or δJ_1 orthogonally onto $\mathrm{aff}(S^n)$ rather than use $K_2(m)$ and $J_2(m)$. These results and computational experience reinforcing them are reported in [67].

<u>XI.4 The Directional Density for Triangulations with Crogs</u>. Let F_i, $i = 0,1$, be as in 2.1, and apply a homotopy algorithm with artificial labelling only on level 0. Then it is easy to see that the simplices generated are those meeting $[(\frac{1}{2},c^{1T})^T, (1,c^{0T})^T] \cup (0,\frac{1}{2}] \times \{c^1\}$. In general, more complicated paths of simplices arise. Our measures are designed to predict the performance of a triangulation when the simplices

(i) lie close to a "horizontal" straight line in one layer $[2^{-k-1},2^{-k}] \times R^n$; or

(ii) lie close to an "angled" straight line moving downwards (or upwards), with perhaps some horizontal drift.

The measures are appropriate for triangulations for which each triangulated layer resembles 2^{-k} times the triangulated layer $[\frac{1}{2},1] \times R^n$, for example, J_3 and K_3.

4.1 Definition. Let G be a triangulation with continuous refinement of grid size.

(a) For $\frac{1}{2} < h < 1$, let $x' = (h,x^T)^T$ and $d' = (0,d^T)^T$ for $x,d \in R^n$. Define $N(G,x',d',\lambda)$ analogously to 3.1 and let $HN(G,x,h,d,\lambda) = N(G,x',d',\lambda)$. Define $HN(G,h,d)$ and $HN(G,h)$ by averaging and taking limits with respect to x, λ, and d as in 3.1.

(b) For $0 < k \in Z$ and $x,d \in R^n$ let $x' = (1,x^T)^T$ and $d' = (-1,d^T)^T$. Let $VN(G,x,d,k)$ denote the number of simplices of G met by $[x', x' + (1-2^{-k})d']$ divided by k. Let $VN(G,d,k)$ denote the limit as $\rho \to \infty$ of the average of $VN(G,x,d,k)$ for x uniformly distributed in $B(0,\rho)$. Let $VN(G,d)$ be the limit of $VN(G,d,k)$ as $k \to \infty$.

We call $HN(G,h,d)$ and $HN(G,h)$ the horizontal directional density of G at height h in direction d and the average horizontal directional density of G at height h. We call $VN(G,d)$ the vertical directional density of G with drift d.

Note the very important point that none of the measures except $VN(G,0)$ corresponds to minimal paths. None except $VN(G,d)$ and $VN(G,0)$ corresponds to paths realized with an affine function. Rather, the measures merely give some idea of the expected performance of a triangulation with crogs.

As in section 3, we can average over x in $[0,2]^n$ for J_3 or $[0,1]^n$ for K_3 rather than $B(0,\rho)$. We can also count facets crossed rather than $(n+1)$-simplices met.

We therefore need to characterize points in facets of J_3 and K_3. Let $\beta(\lambda)$ denote the distance from λ to a closest odd integer.

4.2 Lemma. Let $x \in (0,1] \times R^n$ with $2^{-k+1} \geq x_0 \geq 2^{-k}$, and define $y_0 = 2^{-k}$. Then x lies in a facet of J_3 iff

(a) $x_i/2y_0 \in Z$ for $i \in N_0$;

(b) $(x_i \pm x_0)/2y_0 \in Z$ for $i \in N$;

(c) $(x_i \pm x_j)/2y_0 \in Z$ and $\beta(x_i/y_0) > x_0/y_0 - 1$ for $i < j \in N$; or

(d) $(x_i \pm x_j)/4y_0 \in Z$ and $\beta(x_i/y_0) < x_0/y_0 - 1$ for $i < j \in N$.

Proof. The result follows easily from (*) in the proof of X.2.4.

Examples $(n = 2)$. See X.2.3.

Then AHI and IJK are facets of type (a), OBJ and OBI are facets of type (b), OAI and OCJ are facets of type (c), and OIK is a facet of type (d). A facet of type (d) with the positive sign appears in $[\frac{1}{2},1] \times [1,2] \times [0,1]$, for example.

4.3 Lemma. Let $x \in (0,1] \times R^n$ with $2^{-k+1} \geq x_0 \geq 2^{-k}$, and define $y_0 = 2^{-k}$. Then x lies in a facet of K_3 iff

(a) $x_i/2y_0 \in Z$ for $i \in N_0$;

(b) $(x_i \pm x_0)/2y_0 \in Z$ for $i \in N$;

(c) $(x_i - x_j)/2y_0 \in Z$ for $i < j \in N$; or

(d) $(x_i - x_j + x_0)/2y_0 \in Z$, $\lfloor x_i/y_0 \rfloor$ is even and $\beta(x_i/y_0) > x_0/y_0 - 1$ for $i \neq j \in N$.

(Note that (d) is asymmetric in i and j; recall $\lfloor \lambda \rfloor$ is the greatest integer not greater than λ.)

Proof. The result follows from (**) following X.3.4.

Examples $(n = 2)$. See X.3.3.

Then AHI and IJK are facets of type (a), OBJ and OBI are facets of type (b), OAI and OIK are facets of type (c), and BDJ and FHL are facets of type (d).

Then the following result holds.

4.4 Theorem.

(a) $HN(J_3,h,d) = 3 \sum_i |d_i| + (\frac{3}{2} - h) \sum_{i<j} (|d_i+d_j| + |d_i-d_j|)$.

(b) $HN(K_3,h,d) = 3 \sum_i |d_i| + (3-2h) \sum_{i<j} |d_i-d_j|$.

(c) With g_n defined as in 3.2, $HN(J_3,h) = HN(K_3,h) = (3n + (3-2h)\binom{n}{2}\sqrt{2})g_n$.

(d) $VN(J_3,d) = 1 + \sum_i \frac{1}{2}(|d_i| + |d_i+1| + |d_i-1|) + \frac{3}{8} \sum_{i<j} (|d_i+d_j| + |d_i-d_j|)$.

(e) $VN(K_3,d) = 1 + \sum_i \frac{1}{2}(|d_i| + |d_i+1| + |d_i-1|)$
$$+ \sum_{i<j} (\tfrac{1}{2}|d_i-d_j| + \tfrac{1}{8}|d_i-d_j+1| + \tfrac{1}{8}|d_i-d_j-1|).$$
[All indices of sums run over N.]

Proof. See [65]. The proof is similar to that of 3.2, but one must take into consideration the side conditions in facets of types (c) and (d) of 4.2 and (d) of 4.3. □

It is instructive to note where the various terms above originate. For the horizontal directional density, the first terms come from facets of types (a) and (b) in 4.2 and 4.3. Facets of types (c) and (d) of 4.2 contribute $2-2h$ and $h-\frac{1}{2}$ of the sums over $i<j$. Similarly facets of types (c) and (d) of 4.3 contribute 1 and $2-2h$ of the sum over $i<j$.

For the vertical directional densities, the first two terms in both cases come from facets of types (a) and (b) in 4.2 and 4.3. For J_3, facets of type (c) contribute $\frac{1}{4}$ of the sum over $i<j$ and those of type (d) contribute $\frac{1}{8}$ of the sum. For K_3, the first part of the sum over $i<j$ comes from facets of type (c) in 4.3, while the remaining parts come from facets of type (d).

Comparing J_3 and K_3 in 4.4, we conclude that their horizontal measures are similar. K_3 is more efficient if we know the direction to be followed, and J_3 is less direction-dependent. If the terms $\frac{1}{8}|d_i-d_j+1|$ and $\frac{1}{8}|d_i-d_j-1|$ were $\frac{1}{8}|d_i-d_j|$ and $\frac{1}{8}|d_i+d_j|$, we could make the same statement about the vertical measures. However, the extra +1 or -1 means K_3 is inferior when $\|d\|$ is small. In particular, for $d = 0$ we have $VN(J_3,0) = n+1$ and $VN(K_3,0) = n + 1 + \frac{1}{4}\binom{n}{2}$.

From the analysis above, it is clear that the discrepancy between J_3 and K_3 is caused by facets of type (d) in 4.3. Indeed, it is the dependence of these facets on x_0 that causes problems. A vertical path through one layer of J_3 or K_3 always meets one facet of type (a) $(i = 0)$ and n of type (b). For K_3, it also meets some facets of type (d) unless x is suitably chosen. Using one layer of K_3 in a restart algorithm circumvents the problem since with $x = \frac{1}{2}(n+1)^{-1}(n,n-1,\ldots,1)^T$, $[\frac{1}{2},1] \times \{x\}$ meets only $n+1$ simplices.

REFERENCES

1. Allgower, E. L., "Application of a Fixed Point Search Algorithms to Nonlinear Problems having Several Solutions," in Fixed Points: Algorithms and Applications, S. Karamardian (ed.), Academic Press, to appear.

2. Arrow, K. J., and F. H. Hahn, General Competitive Analysis, Holden-Day, San Francisco, 1971.

3. Berge, C., Topological Spaces (translated by E. M. Patterson), MacMillan, New York, 1963.

4. Bohl, P., "Über die Bewegung eines Mechanischen Systems in der Nähe einer Gleichgewichtslage," J. Reine Angew. Math. 127 (1904), 179-276.

5. Brouwer, L. E. J., "Über Abbildung von Mannigfaltigkeiten," Math. Ann. 71 (1910), 97-115.

6. Cohen, D. I. A., "On the Sperner Lemma," J. Comb. Theory 2 (1967), 585-587.

7. Cottle, R. W., "Nonlinear Programs with Positively Bounded Jacobians," J. SIAM Appl. Math. 14 1 (1966), 147-158.

8. Dantzig, G. B., Linear Programming and Extensions, Princeton University Press, Princeton, N. J. (1963), 621 pp.

9. Eaves, B. C., "Nonlinear Programming Via Kakutani Fixed Points," Working Paper No. 294, Center for Research in Management Science, University of California, Berkeley (1970).

10. ______ "An Odd Theorem," Proc. AMS 26 (1970), 509-513.

11. ______ "On the Basic Theory of Complementarity," Mathematical Programming 1 1 (1971), 68-75.

12. ______ "Computing Kakutani Fixed Points," SIAM J. of Appl. Math. 21 2 (1971), 236-244.

13. ______ "The Linear Complementarity Problem," Management Science 17 9 (1971), 612-634.

14. ______ "Homotopies for Computation of Fixed Points," Mathematical Programming 3 1 (1972), 1-22.

15. ______ "A Short Course in Solving Equations with PL Homotopies," Department of Operations Research, Stanford University (September 1975).

16. Eaves, B. C., and R. Saigal, "Homotopies for Computation of Fixed Points on Unbounded Regions," Mathematical Programming 3 2 (1972), 225-237.

17. Eaves, B. C., and H. Scarf, "The Solution of Systems of Piecewise Linear Equations," Cowles Foundation Discussion Paper, No. 390, Yale University (1975), 67 pp.

18. Fan, K., "Simplicial Maps from an Orientable n-Pseudomanifold into S^m with the Octahedral Triangulation," J. Comb. Theory 2 4 (1967), 588-602.

19. Fisher, M. L., and F. J. Gould, "A Simplicial Algorithm for the Nonlinear Complementarity Problem," Mathematical Programming 6 3 (1974), 281-300.

20. Fisher, M. L., F. J. Gould, and J. W. Tolle, "A New Simplicial Approximation Algorithm with Restarts: Relations Between Convergence and Labelling," in Fixed Points: Algorithms and Applications, S. Karamardian (ed.), Academic Press, to appear.

21. Freidenfelds, J., "Fixed-Point Algorithms and Almost-Complementary Sets," TR 71-17, Operations Research House, Stanford University (1971).

22. ______ "A Set Intersection Theorem and Applications," Mathematical Programming 7 2 (1974), 199-211.

23. Freudenthal, H., "Simplizialzerlegungen von Beschränkter Flachheit," Annals of Mathematics 43 3 (1942), 580-582.

24. Garcia, C. B., C. E. Lemke, and H. Luethi, "Simplicial Approximation of an Equilibrium Point for Non-Cooperative N-Person Games," Mathematical Programming, Ed: T. C. Hu and S. M. Robinson, Academic Press, New York-London (1973), 227-260.

25. Gochet, W., E. Loute, and D. Solow, "Comparative Computer Results of Three Algorithms for Solving Prototype Geometric Programming Problems," CORE Reprint 212, Belgium (1974).

26. Gould, F. J., and J. W. Tolle, "A Unified Approach to Complementarity in Optimization," Discrete Math. 7 (1974), 225-271.

27. ______ "Finite and Constructive Conditions for a Solution to $f(x) = 0$," Center for Mathematical Studies in Business and Economics Report 7515, University of Chicago (March 1975).

28. Hansen, T., and T. C. Koopmans, "On the Definition and Computation of a Capital Stock Invariant under Optimization," J. Economic Theory 5 3 (1972), 487-523.

29. Hansen, T., and H. Scarf, "On the Applications of a Recent Combinatorial Algorithm," Cowles Foundation Discussion Paper No. 272, Yale University (1969).

30. Hirsch, M. W., "A Proof of the Nonretractibility of a Cell onto Its Boundary," Proc. of AMS 14 (1963), 364-365.

31. Jeppson, M. M., "A Search for the Fixed Points of a Continuous Mapping," Mathematical Topics in Economic Theory and Computation, Ed: R. H. Day and S. M. Robinson (1972), 122-129.

32. Kakutani, S., "A Generalization of Brouwer's Fixed Point Theorem," Duke Math. J. 8 (1941), 457-459.

33. Karamardian, S., "The Complementarity Problem," Mathematical Programming 2 (1972), 107-129.

34. ______ (Editor), Fixed Points: Algorithms and Applications, Proc. Conf. on Computing Fixed Points with Applications, Clemson University, Clemson, South Carolina (1975).

35. Kellogg, R. B., T. Y. Li, and J. Yorke, "A Constructive Proof of the Brouwer Fixed Point Theorem and Computational Results," unpublished paper, University of Maryland and University of Utah (1975), 20 pp.

36. Knaster, B., C. Kuratowski, and S. Mazurkiewicz, "Ein Beweis des Fixpunktsatzes für n-dimensionale Simplexe," Fund. Math. 14 (1929), 132-137.

37. Kuhn, H. W., "Some Combinatorial Lemmas in Topology," IBM J. Research and Develop. 4 5 (1960), 518-524.

38. Kuhn, H. W., "Simplicial Approximation of Fixed Points," Proc. Nat. Acad. Sci., U.S.A. 61 (1968), 1238-1242.

39. _____ "Approximate Search for Fixed Points," in: Computing Methods in Optimization Problems - 2, Academic Press, New York, 1969.

40. _____ "How to Compute Economic Equilibria by Pivotal Methods," Department of Economics and Mathematics, Princeton University (1975), 25 pp.

41. Kuhn, H. W., and J. G. MacKinnon, "The Sandwich Method for Finding Fixed Points," J. Optimization Theory and Applications 17 (1975), 189-204.

42. Lefschetz, S., Introduction to Topology, Princeton University Press, Princeton, 1949.

43. Lemke, C. E., "Recent Results on Complementarity Problems," Nonlinear Programming, Ed: J. B. Rosen, O. L. Mangasarian, and K. Ritter, Academic Press, New York (1970), 349-384.

44. Lemke, C. E., and S. J. Grotzinger, "On Generalizing Shapley's Index Theory to Labelled Pseudo Manifolds," Rensselaer Polytechnic Institute (1974).

45. Lemke, C. E., and J. T. Howson, Jr., "Equilibrium Points of Bimatrix Games," SIAM J. on Appl. Math. 12 2 (1964), 413-423.

46. Lyusternik, L. A., Convex Figures and Polyhedra (translated by T. Jefferson Smith), Dover Press, New York, 1963.

47. Mara, P. S., "Triangulations of a Cube," M.S. Thesis, Colorado State University, Fort Collins, Colorado (1972).

48. Merrill, O. H., "Applications and Extensions of an Algorithm that Computes Fixed Points of Certain Non-Empty, Convex, Upper Semi-Continuous Point to Set Mappings," TR 71-7, Department of Industrial Engineering, University of Michigan (1971).

49. _____ "Applications and Extensions of an Algorithm that Computes Fixed Points of Certain Upper Semi-Continuous Point to Set Mappings, Ph.D. Dissertation, Dept. of Ind. Engineering, University of Michigan (1972), 228 pp.

50. Moré, J., "Coercivity Conditions in Nonlinear Complementarity Problems," SIAM Review 16 1 (1974), 15 pp.

51. Rockafellar, R. T., Convex Analysis, Princeton University Press, Princeton, 1970.

52. Saigal, R., "Investigations into the Efficiency of Fixed Point Algorithms," in Fixed Points: Algorithms and Applications, S. Karamardian (ed.), Academic Press, to appear.

53. _____ "On Paths Generated by Fixed Point Algorithms," in preparation.

54. _____ "On the Convergence Rate of Algorithms for Solving Equations that are Based on Complementarity Pivoting," in preparation.

55. Saigal, R., D. Solow, and L. A. Wolsey, "A Comparative Study of Two Algorithms to Compute Fixed Points over Unbounded Regions," presented at the 8th Mathematical Programming Symposium, Stanford (1975).

56. Scarf, H., "The Core of an N Person Game," Econometrica 35 1 (1967), 50-69.

57. Scarf, H., "The Approximation of Fixed Points of a Continuous Mapping, SIAM J. Appl. Math. 15 5 (1967), 1328-1343.

58. Scarf, H. E., and T. Hansen, Computation of Economic Equilibria, Yale University Press, New Haven (1973) 249 pp.

59. Shapley, L. S., "On Balanced Games Without Side Payments," Mathematical Programming, Ed: T. C. Hu and S. M. Robinson, Academic Press, New York-London (1973), 261-290.

60. ______ "A Note on the Lemke-Howson Algorithm," Mathematical Programming Study 1 (1974), 175-189.

61. Smart, D. R., Fixed Point Theorems, Cambridge University Press, Cambridge, 1974.

62. Sperner, E., "Neuer Beweis für die Invarianz der Dimensionszahl und des Gebietes," Abh. Math. Sem. Univ. Hamburg 6.

63. Todd, M. J., "A Generalized Complementary Pivoting Algorithm," Mathematical Programming 6 3 (1974), 243-263.

64. ______ "Union Jack Triangulations," TR 220, Dept. of Operations Research, Cornell University, Fixed Points: Algorithms and Applications, Ed: Stepan Karamardian, Academic Press, (1975).

65. ______ "On Triangulations for Computing Fixed Points," TR 234, Department of Operations Research, Cornell University (1974), 33 pp.

66. ______ "Orientation in Complementary Pivot Algorithms," TR 249, Department of Operations Research, Cornell University (1975), 22 pp.

67. ______ "Improving the Convergence of Fixed-Point Algorithms," TR 276, Department of Operations Research, Cornell University (October 1975).

68. Tompkins, C. B., "Sperner's Lemma and Some Extensions," in: E. F. Beckenbach (ed.), Applied Combinatorial Mathematics, Wiley, New York, 1964.

69. Tucker, A. W., "Some Topological Properties of Disk and Sphere," Proc. First Canadian Math. Congress (1945), 285-309.

70. Uzawa, H., "Walras' Existence Theorem and Brouwer's Fixed-Point Theorem," Economic Studies Quarterly 13, 1.

71. Vertgeim, B. A., "On an Approximate Determination of the Fixed Points of Continuous Mappings," Soviet Math. Dokl. 11 (1970), 295-298.

72. Whitney, H., Geometric Integration Theory, Princeton University Press, Princeton, 1957.

73. Wilmuth, R. J., "The Computations of Fixed Points," Department of Operations Research, Stanford University, Ph.D. Thesis (1973).

74. Wolsey, L. A., "Convergence, Simplicial Paths and Acceleration Methods for Simplicial Approximation Algorithms for Finding a Zero of a System of Nonlinear Equations," CORE Discussion Paper No. 7427, Belgium (1974).

Vol. 59: J. A. Hanson, Growth in Open Economies. V, 128 pages. 1971.

Vol. 60: H. Hauptmann, Schätz- und Kontrolltheorie in stetigen dynamischen Wirtschaftsmodellen. V, 104 Seiten. 1971.

Vol. 61: K. H. F. Meyer, Wartesysteme mit variabler Bearbeitungsrate. VII, 314 Seiten. 1971.

Vol. 62: W. Krelle u. G. Gabisch unter Mitarbeit von J. Burgermeister, Wachstumstheorie. VII, 223 Seiten. 1972.

Vol. 63: J. Kohlas, Monte Carlo Simulation im Operations Research. VI, 162 Seiten. 1972.

Vol. 64: P. Gessner u. K. Spremann, Optimierung in Funktionenräumen. IV, 120 Seiten. 1972.

Vol. 65: W. Everling, Exercises in Computer Systems Analysis. VIII, 184 pages. 1972.

Vol. 66: F. Bauer, P. Garabedian and D. Korn, Supercritical Wing Sections. V, 211 pages. 1972.

Vol. 67: I. V. Girsanov, Lectures on Mathematical Theory of Extremum Problems. V, 136 pages. 1972.

Vol. 68: J. Loeckx, Computability and Decidability. An Introduction for Students of Computer Science. VI, 76 pages. 1972.

Vol. 69: S. Ashour, Sequencing Theory. V, 133 pages. 1972.

Vol. 70: J. P. Brown, The Economic Effects of Floods. Investigations of a Stochastic Model of Rational Investment. Behavior in the Face of Floods. V, 87 pages. 1972.

Vol. 71: R. Henn und O. Opitz, Konsum- und Produktionstheorie II. V, 134 Seiten. 1972.

Vol. 72: T. P. Bagchi and J. G. C. Templeton, Numerical Methods in Markov Chains and Bulk Queues. XI, 89 pages. 1972.

Vol. 73: H. Kiendl, Suboptimale Regler mit abschnittweise linearer Struktur. VI, 146 Seiten. 1972.

Vol. 74: F. Pokropp, Aggregation von Produktionsfunktionen. VI, 107 Seiten. 1972.

Vol. 75: GI-Gesellschaft für Informatik e.V. Bericht Nr. 3. 1. Fachtagung über Programmiersprachen · München, 9.–11. März 1971. Herausgegeben im Auftrag der Gesellschaft für Informatik von H. Langmaack und M. Paul. VII, 280 Seiten. 1972.

Vol. 76: G. Fandel, Optimale Entscheidung bei mehrfacher Zielsetzung. II, 121 Seiten. 1972.

Vol. 77: A. Auslender, Problèmes de Minimax via l'Analyse Convexe et les Inégalités Variationelles: Théorie et Algorithmes. VII, 132 pages. 1972.

Vol. 78: GI-Gesellschaft für Informatik e.V. 2. Jahrestagung, Karlsruhe, 2.–4. Oktober 1972. Herausgegeben im Auftrag der Gesellschaft für Informatik von P. Deussen. XI, 576 Seiten. 1973.

Vol. 79: A. Berman, Cones, Matrices and Mathematical Programming. V, 96 pages. 1973.

Vol. 80: International Seminar on Trends in Mathematical Modelling, Venice, 13–18 December 1971. Edited by N. Hawkes. VI, 288 pages. 1973.

Vol. 81: Advanced Course on Software Engineering. Edited by F. L. Bauer. XII, 545 pages. 1973.

Vol. 82: R. Saeks, Resolution Space, Operators and Systems. X, 267 pages. 1973.

Vol. 83: NTG/GI-Gesellschaft für Informatik, Nachrichtentechnische Gesellschaft. Fachtagung „Cognitive Verfahren und Systeme", Hamburg, 11.–13. April 1973. Herausgegeben im Auftrag der NTG/GI von Th. Einsele, W. Giloi und H.-H. Nagel. VIII, 373 Seiten. 1973.

Vol. 84: A. V. Balakrishnan, Stochastic Differential Systems I. Filtering and Control. A Function Space Approach. V, 252 pages. 1973.

Vol. 85: T. Page, Economics of Involuntary Transfers: A Unified Approach to Pollution and Congestion Externalities. XI, 159 pages. 1973.

Vol. 86: Symposium on the Theory of Scheduling and its Applications. Edited by S. E. Elmaghraby. VIII, 437 pages. 1973.

Vol. 87: G. F. Newell, Approximate Stochastic Behavior of n-Server Service Systems with Large n. VII, 118 pages. 1973.

Vol. 88: H. Steckhan, Güterströme in Netzen. VII, 134 Seiten. 1973.

Vol. 89: J. P. Wallace and A. Sherret, Estimation of Product. Attributes and Their Importances. V, 94 pages. 1973.

Vol. 90: J.-F. Richard, Posterior and Predictive Densities for Simultaneous Equation Models. VI, 226 pages. 1973.

Vol. 91: Th. Marschak and R. Selten, General Equilibrium with Price-Making Firms. XI, 246 pages. 1974.

Vol. 92: E. Dierker, Topological Methods in Walrasian Economics. IV, 130 pages. 1974.

Vol. 93: 4th IFAC/IFIP International Conference on Digital Computer Applications to Process Control, Part I. Zürich/Switzerland, March 19–22, 1974. Edited by M. Mansour and W. Schaufelberger. XVIII, 544 pages. 1974.

Vol. 94: 4th IFAC/IFIP International Conference on Digital Computer Applications to Process Control, Part II. Zürich/Switzerland, March 19–22, 1974. Edited by M. Mansour and W. Schaufelberger. XVIII, 546 pages. 1974.

Vol. 95: M. Zeleny, Linear Multiobjective Programming. X, 220 pages. 1974.

Vol. 96: O. Moeschlin, Zur Theorie von Neumannscher Wachstumsmodelle. XI, 115 Seiten. 1974.

Vol. 97: G. Schmidt, Über die Stabilität des einfachen Bedienungskanals. VII, 147 Seiten. 1974.

Vol. 98: Mathematical Methods in Queueing Theory. Proceedings 1973. Edited by A. B. Clarke. VII, 374 pages. 1974.

Vol. 99: Production Theory. Edited by W. Eichhorn, R. Henn, O. Opitz, and R. W. Shephard. VIII, 386 pages. 1974.

Vol. 100: B. S. Duran and P. L. Odell, Cluster Analysis. A Survey. VI, 137 pages. 1974.

Vol. 101: W. M. Wonham, Linear Multivariable Control. A Geometric Approach. X, 344 pages. 1974.

Vol. 102: Analyse Convexe et Ses Applications. Comptes Rendus, Janvier 1974. Edited by J.-P. Aubin. IV, 244 pages. 1974.

Vol. 103: D. E. Boyce, A. Farhi, R. Weischedel, Optimal Subset Selection. Multiple Regression, Interdependence and Optimal Network Algorithms. XIII, 187 pages. 1974.

Vol. 104: S. Fujino, A Neo-Keynesian Theory of Inflation and Economic Growth. V, 96 pages. 1974.

Vol. 105: Optimal Control Theory and its Applications. Part I. Proceedings 1973. Edited by B. J. Kirby. VI, 425 pages. 1974.

Vol. 106: Optimal Control Theory and its Applications. Part II. Proceedings 1973. Edited by B. J. Kirby. VI, 403 pages. 1974.

Vol. 107: Control Theory, Numerical Methods and Computer Systems Modeling. International Symposium, Rocquencourt, June 17–21, 1974. Edited by A. Bensoussan and J. L. Lions. VIII, 757 pages. 1975.

Vol. 108: F. Bauer et al., Supercritical Wing Sections II. A Handbook. V, 296 pages. 1975.

Vol. 109: R. von Randow, Introduction to the Theory of Matroids. IX, 102 pages. 1975.

Vol. 110: C. Striebel, Optimal Control of Discrete Time Stochastic Systems. III. 208 pages. 1975.

Vol. 111: Variable Structure Systems with Application to Economics and Biology. Proceedings 1974. Edited by A. Ruberti and R. R. Mohler. VI, 321 pages. 1975.

Vol. 112: J. Wilhelm, Objectives and Multi-Objective Decision Making Under Uncertainty. IV, 111 pages. 1975.

Vol. 113: G. A. Aschinger, Stabilitätsaussagen über Klassen von Matrizen mit verschwindenden Zeilensummen. V, 102 Seiten. 1975.

Vol. 114: G. Uebe, Produktionstheorie. XVII, 301 Seiten. 1976.

Vol. 115: Anderson et al., Foundations of System Theory: Finitary and Infinitary Conditions. VII, 93 pages. 1976

Vol. 116: K. Miyazawa, Input-Output Analysis and the Structure of Income Distribution. IX, 135 pages. 1976.

Vol. 117: Optimization and Operations Research. Proceedings 1975. Edited by W. Oettli and K. Ritter. IV, 316 pages. 1976.

Vol. 118: Traffic Equilibrium Methods, Proceedings 1974. Edited by M. A. Florian. XXIII, 432 pages. 1976.

Vol. 119: Inflation in Small Countries. Proceedings 1974. Edited by H. Frisch. VI, 356 pages. 1976.

Vol. 120: G. Hasenkamp, Specification and Estimation of Multiple-Output Production Functions. VII, 151 pages. 1976.

Vol. 121: J. W. Cohen, On Regenerative Processes in Queueing Theory. IX, 93 pages. 1976.

Vol. 122: M. S. Bazaraa, and C. M. Shetty,Foundations of Optimization VI. 193 pages. 1976

Vol. 123: Multiple Criteria Decision Making. Kyoto 1975. Edited by M. Zeleny. XXVII, 345 pages. 1976.

Vol. 124: M. J. Todd. The Computation of Fixed Points and Applications. VII, 129 pages. 1976.

Vol. 125; Karl C. Mosler. Optimale Transportnetze. Zur Bestimmung ihres kostengünstigsten Standorts bei gegebener Nachfrage. VI, 142 Seiten. 1976.

Vol. 126: Energy, Regional Science and Public Policy. Energy and Environment I. Proceedings 1975. Edited by M. Chatterji and P. Van Rompuy. VIII, 316 pages. 1976.

Vol. 127: Environment, Regional Science and Interregional Modeling. Energy and Environment II. Proceedings 1975. Edited by M. Chatterji and P. Van Rompuy. IX, 211 pages. 1976.